EDIBLE & MEDICINAL PLANTS
of the
ROCKIES

Linda Kershaw

LONE PINE

Lone Pine Publishing

The Publisher: Lone Pine Publishing

10145 – 81 Ave.	1901 Raymond Ave. SW, Suite C
Edmonton, AB T6E 1W9	Renton, WA 98055
Canada	USA

Website: http://www.lonepinepublishing.com

Canadian Cataloguing in Publication Data

Kershaw, Linda J., 1951–
  Edible and medicinal plants of the Rockies

  Includes bibliographical references and index.
  ISBN 1-55105-229-6

  1. Plants, Useful—Rocky Mountains Region. 2. Wild plants, Edible—Rocky Mountains Region. 3. Medicinal plants—Rocky Mountains Region. I. Title.
QK98.4.R6K47 2000   581.6'3'0978   C00-910426-7

Editorial Director: Nancy Foulds
Project Editor: Lee Craig
Editorial: Lee Craig, Erin McCloskey
Technical Review: Julie Hrapko
Production Manager: Jody Reekie
Layout & Production: Curt Pillipow, Heather Markham
Book Design: Heather Markham, Rob Weidemann & Curt Pillipow
Cover Design: Rob Weidemann
Production Support: Arlana Anderson-Hale
Cover Photographs: Linda Kershaw
Illustrations: Ian Sheldon (except for glossary illustrations, by Linda Kershaw)
Separations & Film: Elite Lithographers Company

The photographs in this book are reproduced with the generous permission of their copyright holders. A full list of photo credits is given on p. 260.

The following photos were supplied with the kind permission of the Glenbow Archives, Calgary, Canada: NA-7-54 (p. 13); NA-667-486 (p. 14); NA-667-345 (p. 18); NA-2244-35 (p. 19); NA-2244-27 (p. 21).

DISCLAIMER: This guide is not meant to be a 'how-to' reference guide for consuming wild plants. We do not recommend experimentation by readers, and we caution that many of the plants in the Rockies, including some traditional medicines, are poisonous and harmful.

We acknowledge the financial support of the Government of Canada through the Book Publishing Industry Development Program (BPIDP) for our publishing activities.

PC: 4

Contents

The following people are thanked for their valued contributions to this book:

Ian Sheldon, for his delicate and beautiful illustrations;

the many photographers who allowed us to use their incredible photographs, including the archival photographs from the Glenbow Museum;

Julie Hrapko, who provided a helpful review of the text;

Lone Pine's talented editorial and production staff;

and finally, the native peoples, settlers, botanists and writers who kept written records or oral accounts of the many uses of plants in the Rocky Mountains.

PICTORIAL GUIDE

TREES

Firs
p. 26

Spruces
p. 28

Douglas-fir
p. 30

Western red cedar
p. 31

Hemlocks
p. 32

Larches
p. 34

Two-needled pines
p. 36

Ponderosa pine
p. 37

Five-needled pines
p. 38

White birch
p. 39

Balsam poplar
p. 40

Trembling aspen
p. 41

SHRUBS

Common juniper
p. 42

Cedar-like
junipers, p. 44

Western yew
p. 46

Oaks
p. 47

Willows
p. 48

Alders
p. 50

Gooseberries
p. 52

Prickly currants
p. 54

Black & red
currants, p. 56

Maples
p. 58

Bush-cranberries
p. 60

Elderberries
p. 62

Hawthorns
p. 64

PICTORIAL GUIDE

SHRUBS

| Choke cherry p. 66 | Red cherries p. 67 | Oceansprays p. 68 | Saskatoon p. 69 | Thimbleberry p. 70 |

| Raspberries p. 71 | Wild roses p. 72 | Mountain-ashes p. 74 | Oregon-grapes p. 76 |

| Smooth sumac p. 78 | Skunkbush p. 79 | Buckbrushes p. 80 | Buckthorn & Cascara, p. 82 | Red-osier dogwood, p. 83 |

| Lewis' mock-orange, p. 84 | Canada buffaloberry, p. 85 | Silverberry p. 86 | Big sagebrush p. 87 |

| Rabbitbushes p. 88 | Bearberries & Manzanitas, p. 90 | Labrador teas p. 92 | Pipsissewa p. 93 | False-wintergreens, p. 94 |

| Black huckleberry p. 95 | Blueberries p. 96 | Cranberries p. 98 | Devil's club p. 100 |

PICTORIAL GUIDE

HERBS

Fool's onion
p. 101

Wild onions
p. 102

Wild chives
p. 104

Common camas
p. 105

Beargrass
p. 106

Narrow-leaved yucca
p. 107

Mariposa-lilies
p. 108

Fritillaries
p. 109

Lilies
p. 110

Yellow glacier-lily
p. 111

Mountain marsh-
marigold, p. 112

Rocky Mountain
cow-lily, p. 113

Bedstraws
p. 114

Fireweeds
p. 116

Common evening-
primrose, p. 117

Bitterroot
p. 118

Stonecrop & Roseroot
p. 119

Wild mustards
p. 120

Bittercresses
p. 122

Watercress &
Yellowcress, p. 123

Shepherd's-purse
p. 124

Field pennycress
p. 125

Spider-flower
p. 126

Silverweed
p. 127

Strawberries
p. 128

Clovers
p. 130

Alfalfa
p. 132

PICTORIAL GUIDE

HERBS

Common sweet
clover, p. 133

Sweet-vetches
p. 134

Wild licorice
p. 136

Wild ginger
p. 137

Violets
p. 138

Scarlet globemallow
p. 140

St. John's-wort
p. 141

Prickly-pear cacti
p. 142

Perennial eveningstars
p. 144

Dogbanes
p. 145

Showy milkweed
p. 146

Monument plant
p. 147

Gentians
p. 148

Western blue flax
p. 150

Skullcaps
p. 151

Swamp hedge-nettle
p. 152

Giant-hyssops
p. 153

Wild mints
p. 154

Wild bergamot
p. 156

Great mullein
p. 157

Common hound's-
tongue, p. 158

Gromwells
p. 159

Bunchberry
p. 160

Agoseris
p. 161

Common
dandelion, p. 162

Salsifies
p. 164

Sow thistles
p. 166

PICTORIAL GUIDE

HERBS

Thistles
p. 168

Burdocks
p. 170

Wild sages
p. 172

Chicory
p. 174

Yarrow
p. 175

Pearly everlasting
p. 176

Pineapple-weed
p. 177

Common tansy
p. 178

Tall coneflower &
Black-eyed Susan, p. 179

Common
sunflower, p. 180

Arrow-leaved
balsamroot, p. 181

Coltsfoot
p. 182

Goldenrods
p. 184

Curly-cup
gumweed, p. 186

Wild sarsaparilla
p. 187

Valerians
p. 188

Water-parsnip
p. 190

Cow-parsnip
p. 191

Angelicas
p. 192

Canby's lovage
p. 194

Gairdner's
yampah, p. 195

Desert-parsleys
p. 196

Snakeroot
p. 198

Common
plantain, p. 199

Stinging nettle
p. 200

Strawberry-blite
p. 202

Lamb's-quarters
p. 203

PICTORIAL GUIDE

HERBS

Arrowheads
p. 204

Broad-leaved water-
plantain, p. 205

Bistorts
p. 206

Knotweeds
p. 208

Sheep sorrel
p. 209

Docks
p. 210

Mountain sorrel
p. 212

Bulrushes
p. 213

Cattails
p. 214

Grasses
p. 216

Common
sweetgrass, p. 218

Western
polypody, p. 219

Woodferns
p. 220

Bracken
p. 221

Horsetails & Scouring-
rushes, p. 222

Clubmosses
p. 224

MOSSES & LICHENS

Peat mosses
p. 226

Iceland-lichen
p. 228

Reindeer-lichens
p. 230

Hair lichens
p. 232

POISONOUS PLANTS

Poison-ivy
p. 237

Small bog-laurel
p. 237

False azalea
p. 238

White
rhododendron, p. 238

Common
snowberry, p. 239

PICTORIAL GUIDE

POISONOUS PLANTS

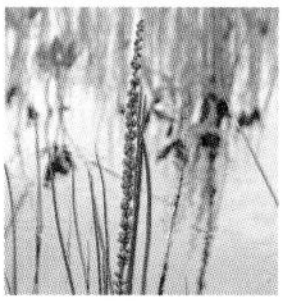
Seaside arrow-grass, p. 239

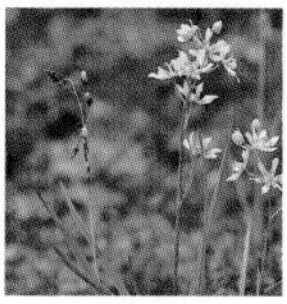
Death-camases p. 240

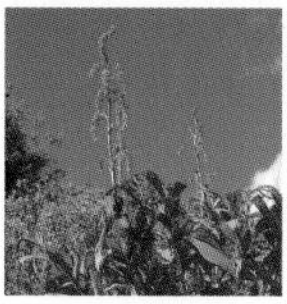
Green false-hellebore, p. 240

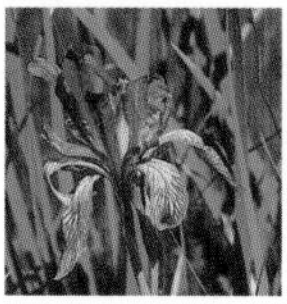
Western blue flag p. 241

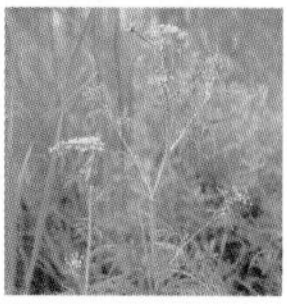
Water-hemlocks p. 241

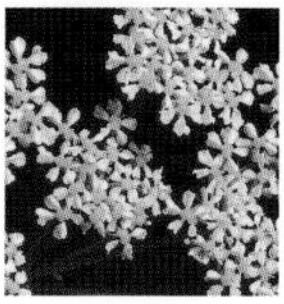
Poison-hemlock p. 242

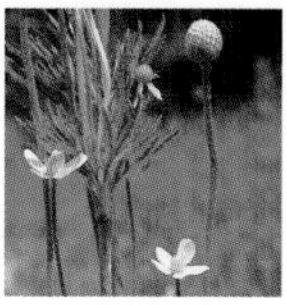
Anemones p. 242

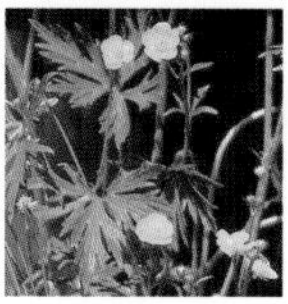
Buttercups p. 243

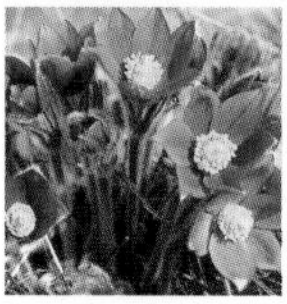
Pasqueflowers p. 244

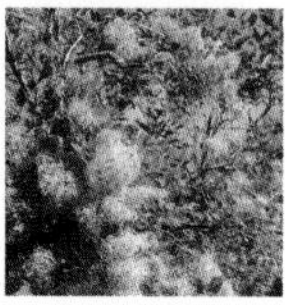
Virgin's-bowers p. 244

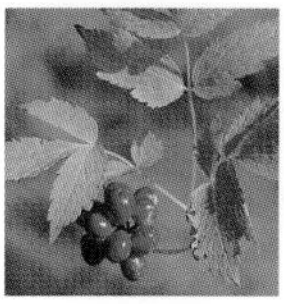
Baneberry p. 245

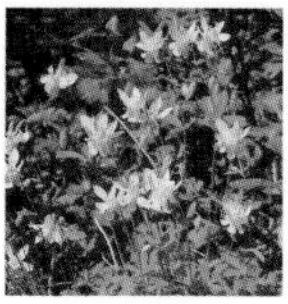
Columbines p. 245

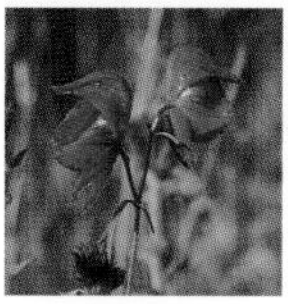
Monkshoods p. 246

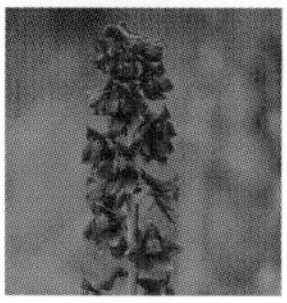
Larkspurs p. 246

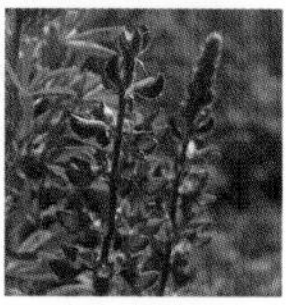
Lupines p. 247

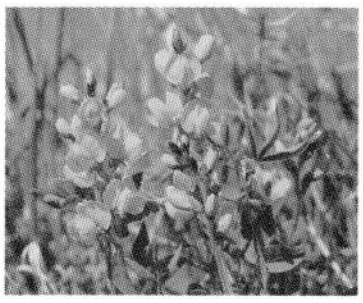
Goldenbeans p. 248

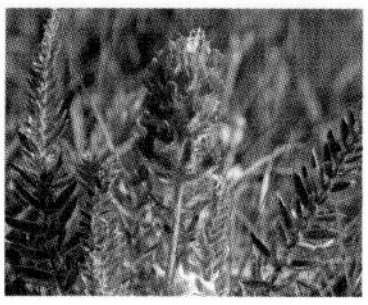
Locoweeds p. 248

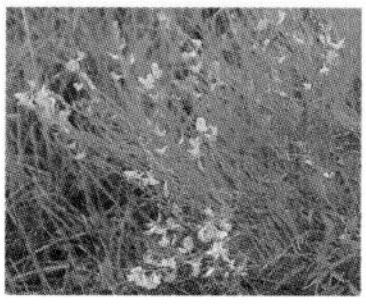
Timber milk-vetch, p. 249

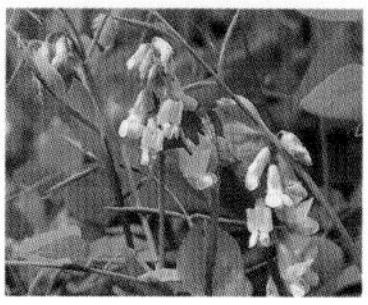
Peavines p. 249

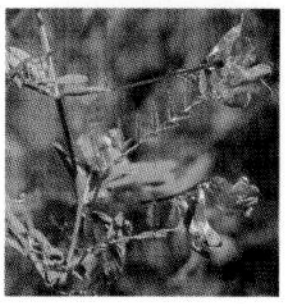
American vetch p. 250

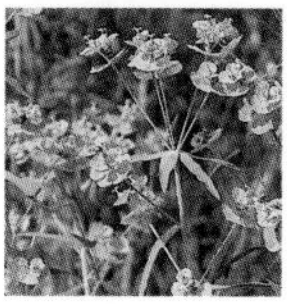
Leafy spurge p. 250

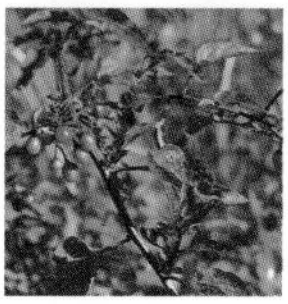
European bittersweet, p. 251

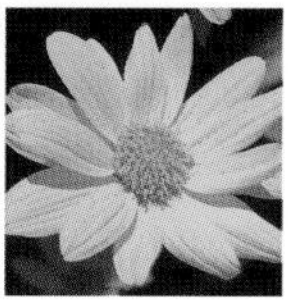
Arnicas p. 251

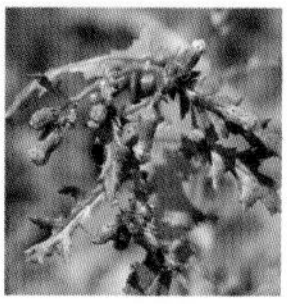
Groundsels p. 252

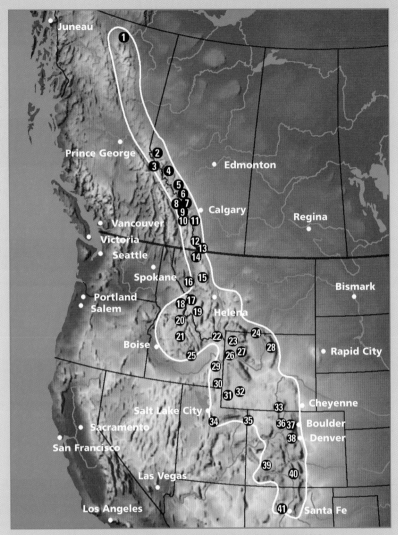

PROTECTED AREAS IN THE ROCKY MOUNTAINS

1 The Muskwa
2 Willmore Wilderness Park
3 Mount Robson & Rearguard Falls Provincial Parks
4 Jasper National Park
5 White Goat Wilderness Area
6 Siffleur Wilderness Area
7 Banff National Park
8 Yoho National Park
9 Mount Assiniboine Provincial Park
10 Kootenay National Park
11 Kananaskis Country & Peter Lougheed Provincial Park
12 Castle Crown Wilderness
13 Waterton Lakes National Park
14 Glacier National Park
15 Great Bear, Bob Marshall & Scapegoat Wilderness Areas
16 National Bison Range
17 Bitterroot Watchable Wildlife Triangle
18 Selway-Bitterroot Wilderness Area
19 Anaconda-Pintler Wilderness
20 Frank Church–River of No Return Wilderness

21 Sawtooth National Recreation Area, Sawtooth Wilderness Area & Land of the Yankee Fork State Park
22 Red Rock Lakes National Wildlife Refuge
23 Yellowstone National Park
24 Pryor Mountain Wild Horse Range
25 Craters of the Moon National Monument
26 Grand Teton National Park
27 Teton & Gros Ventre Wildernesses
28 Cloud Peak Wilderness
29 Grays Lake National Wildlife Refuge
30 Bear Lake National Wildlife Refuge
31 Fossil Butte National Monument
32 Seedskadee National Wildlife Refuge
33 Encampment Wildernesses
34 Timpanogos Cave National Monument
35 Dinosaur National Monument
36 Arapaho National Wildlife Refuge & Illinois River Moose Viewing Site
37 Rocky Mountain National Park
38 Golden Gate Canyon State Park
39 Big Blue & Powderhorn Wilderness Areas
40 Great Sand Dunes National Monument
41 Bandelier National Monument

Introduction

All animals, including humans, depend on plants for survival. Throughout human history, plants have provided us with food, clothing, medicine and shelter. Our recent ancestors needed to know which of their local plants were edible or poisonous, which could heal or harm, and which could provide materials for making implements, clothing and shelters. Today, many of us spend our lives in artificial environments, isolated from our natural surroundings. Most of life's necessities are mass produced elsewhere and purchased as

Spruce-bark tipi

needed. We no longer forage for food, build our shelters or gather fuel for heating and cooking. Instead, we depend on hundreds of people in our society to supply the knowledge, materials and skills necessary to feed, clothe and house us. As we grow increasingly isolated from our natural surroundings, it is easy to forget that the air we breathe, the food we eat and many of the drugs we use come from plants. Recognizing wild plants and knowing how they have been used in the past increases our appreciation of our environment.

With an ever-growing human population and the introduction of aggressive weed species from around the world, fewer and fewer areas remain where 'natural' plant communities persist. Wild plant communities cannot support the large numbers of people now living on this continent, so vast tracts of land have been converted into cultivated fields, pastures and plantations. The remaining wilderness needs to be treated with care and consideration.

Many of the plants that grow wild in the Rocky Mountains are also found in towns, cities and roadsides across the continent; millions of dollars are spent each year controlling weeds that might be better used for food or medicine. An appreciation of how our ancestors survived through the centuries, using many of the plants that surround us everyday, may help us to bridge the gap between the artificial world in which we live and the natural environment in which we evolved.

The main goal of this book is to illustrate the great variety of ways in which plants have been used by humans. To date, few wild plants have been studied as medicines or foods or for other uses. By conserving the biological diversity in the Rockies and elsewhere, we will have the time needed to further study wild plants. Once a species is lost, it can never be recreated, and we can never learn more about it.

DISCLAIMER

This book provides interesting information about many plants in the Rocky Mountains. It is not intended as a 'how-to' guide for living off the land. Only some of the more common and widely used species in the Rockies are described and discussed. Many edible and poisonous plants are not included. No plant or plant extract should be consumed unless you are certain of its identity and toxicity and of your personal potential for allergic reactions. Self-medication with herbal medicines is often unwise, and wild foods should always be used with **caution**. The author and publisher are not responsible for the actions of the reader.

13

Mrs. Oldman Spotted pounding choke cherries

Sources of Information

The information in this book has been collected from many publications (see references, p. 259) and is not based on personal experimentation. No judgment has been made regarding the validity of most reported uses. Instead, a wide range of information is included for your interest.

Many sources disagree about how a plant should be used. One book says that a plant has been used for food, while another reports that it contains toxins and needs special preparation or perhaps should not be eaten at all. Controversy is even greater over medicinal herbs. One author says that a plant was used to treat diarrhea and dysentery, while another reports that it was used to relieve cold symptoms and soothe sore throats. All reports are included in the text to give some idea of the wide range of uses over the years.

Of course, not all reported uses are valid. Some 'edible' species may have been used for food only during famines, when tastier, more nutritious plants were not available. What won't fatten will fill. Medicinal herbs were not necessarily effective or safe. During the smallpox or measles epidemics that swept across North America, most plants were tried at one time or another as a cure. Also, many 'powerful' medicines are lethal poisons when used improperly. More detailed discussions of herbal medicines and food preparation can be found in some of the books listed in the references section.

Using Wild Plants

With growing populations and dwindling areas of natural vegetation, visitors to wilderness regions can no longer expect to collect wild plants. The use of wild resources must be viewed as a privilege, rather than a right. If everyone decided to 'harvest nature's crop,' many species and many plant communities would soon disappear. No one has seeded or tended these wild plants, so neither 'harvest' nor 'crop' apply, though both terms are often used. Many populations of wild medicinal plants have disappeared as a result of uncontrolled 'harvesting,' especially around densely populated areas. However, many common plants, in the wild and in fields or on roadsides closer to home, can be gathered without threatening natural ecosystems. Gathering wild plants simply requires common sense and respect for the environment.

A Few Gathering Tips

1. Gather only common plants, and never take more than five percent of the population (e.g., one plant in 20, one berry in 20). Even then, a cautious personal quota will still deplete the plants if too many people gather them in one area. Remember, plants growing in harsh mountain environments might not have enough energy to produce flowers and fruits every year. Also, don't forget the other local wildlife. Survival of many animals can depend on access to the roots, shoots and fruits that you are harvesting.

2. Never gather plants from protected and/or heavily used areas such as parks and nature preserves. Doing so is not only wrong, but often it is illegal. Be sure to check the regulations for the area you are visiting.

3. Know which part of the plant you need. It isn't always necessary to kill a plant to use it. Don't pull up an entire plant if you are only going to use the flowers or a few of the younger leaves. Take only what you need and damage the plant as little as possible. If you are gathering bulbs or rootstocks, leave smaller pieces buried in the ground to produce new plants. If you want to grow a plant in your garden, try propagating it from seed, rather than transplanting it from the wild.

4. Take only as much as you need. If you are gathering a plant for food, taste a sample. You may not like it, or the berries at this site may not be as sweet and juicy as the ones you gathered last year. Don't take more than you will use.

Rocky Mountain cow-lily

5. Gather plants only when you are certain of their identity. Many irritating and poisonous plants grow wild in the Rockies, and some of them resemble edible or medicinal species. If you are not positive that you have the right plant, don't use it. It is better to eat nothing at all than to be poisoned.

6. Gather plants carefully. Always take care not to accidentally mix in the leaves or rootstocks of nearby plants, which could be poisonous.

No Two Plants Are the Same

Wild plants are highly variable. No two individuals of a species are identical and many characteristics can vary. Some of the more easily observed characteristics include the color, shape and size of stems, leaves, flowers and fruits. Other less obvious features, such as sweetness, toughness, juiciness and concentrations of toxins or drugs, also vary from one plant to the next.

Many factors control plant characteristics. Time is one of the most obvious. All plants change as they grow and age. Usually, young leaves are the most tender, and mature fruits are the largest and sweetest. Underground structures also change throughout the year. Rootstocks, bulbs, tubers and corms are storage chambers for nutrients. During the growing season, these underground structures become smaller, tougher and spongier, as their supplies are moved up to support rapidly growing stems, leaves, flowers and fruits. In late summer, the roots and underground stems swell once again, as the aging upper plant sends nutrients back down for storage, in preparation for next year's growth. Crisp, firm rootstocks remain swollen with starches and sugars through winter and early spring, often making this time of year the best to gather these parts.

Habitat also has a strong influence on plant growth. The leaves of plants from moist, shady sites are often larger, sweeter and more tender than those of plants on dry, sunny hillsides. Berries may be plump and juicy one year, when shrubs have had plenty of moisture,

Common sunflower

but they can become dry and wizened during a drought. Without the proper nutrients and environmental conditions, plants cannot grow and mature.

Finally, the genetic make-up of a species determines how the plant develops and how it responds to its environment. Wild plant populations tend to be much more variable than domestic crops, partly because of their wide range of habitats, but also because of their greater genetic variability. Humans have been planting and harvesting plants for centuries, repeatedly selecting and breeding plants with the most desirable characteristics. This process has produced many highly productive cultivars—trees with larger, sweeter fruits, potatoes with bigger tubers and sunflowers with larger, oilier seeds. These crop species are more productive, and they also produce a specific product each time they are planted. Wild plants are much less reliable.

Wild species have developed from a broader range of ancestors, growing in many different environments, so their genetic make-up is much more variable than that of domestic cultivars. One population may produce sweet juicy berries, while the berries of another population are small and tart; one plant may have low concentrations of a toxin that is plentiful in its neighbor. This variability makes wild plants much more resilient to change. Although their lack of stability may seem to reduce their value as crop species, it is one of their most valuable features. Domestic crops often have few defenses and must be protected from competition and predation. As fungi, weeds and insects continue to develop immunities to pesticides, we repeatedly return to wild plants for new repellents and, more recently, for pest-resistant genes for our crop plants.

The Species Accounts

This book includes several common plants that have been used by people from ancient times to the present. Species are organized by growth form (i.e., trees, shrubs, herbs, including grasses and ferns, mosses and lichens), with closely related or similar plants grouped together for comparison. A section on poisonous plants is at the end of the book. Many entries focus on a single species, but if several similar species (e.g., several violets, *Viola* spp.) have been used in the same ways, two or more species may be described together.

Plant Names

Both common and scientific names are included for each plant. These names follow the taxonomic treatment adopted in *Plants of the Rocky Mountains* (Kershaw et. al. 1998).

Food

Although many common plants in the Rockies are edible, all are not equally palatable. Present-day diets are much higher in sugar and salt and have much more highly refined ingredients than were found in the foods of our ancestors. Consequently, many of the plants and fruits that were eaten in the past are quite bitter by modern standards, and many fiber-rich wild foods can cause digestive upset, at least at first. When trying a wild food for the first time, it is best to take only a small amount. See if you like it and if it agrees with your system, before you go to the trouble of preparing a larger amount.

Many wild greens are much more nutritious than the domestic vegetables that commonly fill our salad bowls, but just because a plant is 'natural' doesn't mean that it is better. The edible parts of wild plants (e.g., tubers or fruits) are usually much smaller than those of domestic varieties (though some people would argue that a small wild strawberry contains all the flavor of a large domestic one). Many wild plants are more bitter and fibrous than their domestic counterparts. Also, wild species are more variable than cultivated varieties, so you may need to sample as you go, to assess the sweetness or tenderness of the plants you are gathering.

High bush-cranberry

This guide is not a cookbook, and there are no specific recipes in the Food sections. However, some notes on preparation are included, if special techniques have been used to improve the flavor or texture of a food or to reduce its toxicity. Food plants have been prepared and served in many different ways.

Raw plants provide salad greens, garnishes and trail-side nibbles. Sweet, tender parts, such as young leaves and shoots, flowers and fruits, are often eaten raw. Sharp-tasting or peppery plants are usually mixed with milder greens or used as garnishes in sandwiches and salads, but they can also provide a refreshing, thirst-quenching snack along the trail.

When eating raw plants, make sure that they are free of contaminants. This preparation may involve careful peeling and washing. Never eat plants from areas that may have been contaminated with pesticides, engine exhaust, sewage or other pollutants. Always use clean water for washing. If you wouldn't drink the water, don't use it to wash foods that are going to be eaten raw.

Boiling improves the palatability of many wild foods. It breaks down plant fibers to tenderize tough plants, and it can also improve flavor.

Bitter, astringent plants are often cooked in one or more changes of water to remove undesirable, water-soluble compounds. This process involves pouring boiling water over the vegetable in a pot, returning the water to a boil, draining, adding fresh hot water, and bringing the pot to a boil again. Sometimes, a pinch of baking soda is added to the water to tenderize tough plants. Boiling can also remove water-soluble toxins, which are then discarded with the cooking water. The heat from cooking may break down some poisons, making certain toxic foods safe to eat.

Most tender plants are cooked as little as possible if they are going to be served as a hot vegetable, but they can also be simmered for long periods in soups as thickeners. Many cooked greens are chilled, mixed with vinegar, oil and seasonings and used in salads.

Baking and **roasting** can also improve texture and palatability. Often, this type of cooking breaks down relatively indigestible carbohydrates, producing sugars that make the food sweeter and easier to digest.

Pit cooking was used by many tribes. This lengthy process began by digging a pit and lining it with stones that had been heated in a fire. The stones were covered with green

Mrs. Maggie Big Belly drying saskatoon berries

leaves, followed by a layer of the food to be cooked, a layer of leaves and matting and finally about 6" (15 cm) of firmly packed soil. Often, sticks were placed vertically in the pit during the layering and when the pit was complete, they were removed and water was poured down to the hot rocks to steam the contents. The holes were then plugged, and a hot fire was built on top of the pit and maintained until cooking was complete. Cooking lasted from several hours to a few days. Large amounts of food could be cooked in pits for social gatherings or for storage for later use.

Foods with tough, outer coverings were often placed on, or covered with, hot coals and left to roast. The burned outer layers were then peeled away, and the edible inner parts were eaten.

Drying was the traditional way of preserving many plant foods. Roots and berries were cleaned and spread on leaves in well-ventilated areas, often in the sun, where they dried in a few days. Smoke and heat from slow-burning fires often kept insects away and increased the rate of drying. Larger, thicker plant parts were usually sliced to speed the drying process. Sometimes fruits or roots were first mashed and/or boiled into mush and then spread in a thin layer (like fruit leather) or formed into cakes and dried. Leaves were usually dried by hanging bundles of plants in a warm, dry place.

Today, fruits and vegetables can be dried in dehydrators or in ovens with the door propped open a crack to allow ventilation. Ovens must be set at very low heat, or the food may cook and thus change its flavor and texture. Many dried foods, especially berries and other sweet fruits, can be eaten dry, as snacks or trail food, but most dried vegetables are best softened by soaking and/or boiling.

Preserving and **canning** have also been widely used to prepare wild foods for storage. Many fruits, including a large number of berries that are too sour to enjoy fresh, make excellent jams and jellies. Basically, fruits are boiled with sugar and water until they thicken to make jams and then strained to make jelly. Some fruits thicken on their own, but others need additional pectin in order to jell. Even fruits that are low in pectin can provide tasty sauces for desserts, but they usually require large amounts of additional sugar. Pickling is a preserving technique that uses vinegar instead of sugar to maintain the quality of the food. Basically, fruits or vegetables are packed in jars with a mixture of spices, covered with boiling vinegar and left to pickle for several weeks. Most cookbooks have recipes and instructions for canning and for making jams, jellies and pickles.

Flours and **cereals** are prepared from starchy fruits (e.g., grass grains, acorns), rootstocks, tubers, bulbs and corms; this preparation usually involves cleaning and thoroughly drying the plant part and then grinding the product into meal or flour.

Seed cleaning involves removing tough outer parts such as the small bracts that surround grass grains and the thick hulls that enclose acorn nut meats. First, the outer parts must be broken free from the seed. In some cases, gently rubbing the fruits between your hands will break and remove the outer shell, but in other cases it is necessary to crack and lightly grind the fruits. Once the grains or meats have been freed, the mixture is repeatedly tossed into the air, poured back and forth between containers or covered with water, so that breezes or buoyancy can separate the heavy, starchy grains from their lighter outer coverings.

Cleaning underground plant parts, such as rootstocks, usually involves thorough washing and the removal of rootlets and tough outer layers. The cleaned, starchy parts are then dried thoroughly and ground into meal or flour. Sometimes, large, starchy rootstocks are peeled and then broken apart under water to free grains of starch from the tough central fibers. The fibers are then discarded, the heavy starch is allowed to settle, and the water is poured off. This starchy residue can be used immediately as a wet dough, or it can be dried and ground into flour.

Traditionally, grinding involved crushing grains between two rocks, but blenders and electric mills have greatly simplified this task. Starchy meal or flour can be eaten alone or mixed with fats, cooked in water as mush, added to soups and stews as a thickener, or used alone or (usually) with other flours in breads, muffins, pancakes, etc.

Mrs. George Lezard washing bitterroot

Parching can greatly improve the flavor of many grains and nutlets. Slightly roast dried seeds or seed-like fruits at low temperatures until they are lightly browned. The ones with hairs or other coverings that are difficult to remove may be lightly burned to remove their outer layers, while at the same time parching the seed within. Sometimes, parching will split open the tough outer shells of fruits, and in a few cases (e.g., Rocky Mountain cow-lily, p. 113) it pops the seeds, producing 'popcorn.'

Steeping, decocting and **juicing** all produce beverages. Many wild plants can be used to make caffeine-free teas, coffees and cold drinks.

Most teas are made by pouring boiling water over leaves, flowers, bark or fruits and leaving the mixture to steep for five to ten minutes to make an infusion. Sometimes, plants are boiled in water to make decoctions. The strength of the tea increases with the length of steeping or boiling time and with the amount of plant material used.

Coffees are made by drying and roasting rootstocks, grains or nutlets until they are brown and then grinding them into meal or powder. These mixtures can be perked or dripped as is, but often they are mixed with domestic coffee as a 'coffee-stretcher' or flavor-enhancer.

Some teas and coffees make excellent cold drinks. A few are chilled and served plain, but most are sweetened with sugar and/or mixed with ginger, lemon, raspberries or other fruit for flavoring. Delicious cold drinks can also be made by mixing mashed fruits with sugar and soda water. Similarly, sour-tasting leaves may be bruised, soaked in water and sweetened to make a refreshing drink similar to lemonade.

Fermentation can transform a great variety of wild fruits, flowers, roots, leaves and barks into delicious beers, wines and vinegars. Generally, this process requires steeping or boiling selected plant parts in water and straining off the liquid, but sometimes the sweet

Rocky Mountain maple

spring sap of trees is used. The plant extract is sweetened with sugar, yeast is added and the mixture is left to bubble (ferment) in a warm place until all of the sugar has been converted into alcohol and gas. Sometimes, when the fermenting liquid is contaminated with certain bacteria, the alcohol turns into acetic acid and vinegar is produced. The quality of homemade wine, beer and vinegar varies greatly from plant to plant and batch to batch. Success requires knowledge of wine-making techniques and the use of proper equipment, combined with a certain amount of experimentation. Many books are available on this topic.

Sugar extraction can be a long, tedious task. The best-known source of wild sugar is the sweet, watery sap of maple and birch trees. Large quantities of sap (about 10–15 times the final volume, depending on sugar content) must be collected in early spring and boiled for many hours to remove the water and concentrate the sugars. Sometimes sugar from sweet plants was dissolved in water, and the liquid was then boiled like sap to produce syrup. Sugar extraction is a lengthy, laborious task, and it is understandable that the traditional diets of native peoples relied largely on berries and other naturally sweet foods for sweetening.

Oil extraction from plants can be time consuming. Most tribes relied on animal fat, because very few wild plants contain enough oil to merit refining. However, some oil-rich seeds (e.g., sunflower seeds) were boiled in water, and the oil that rose to the surface was skimmed off and stored for later use. Water in narrow containers has a smaller surface area and a thicker layer of oil, which makes skimming easier.

Seasonings are provided by strong-flavored or salt-rich plants that may not be edible by themselves, but that add flavor to other foods when used in small amounts. Plants to be used for seasoning were often dried and stored for later use. Some salt-rich plants were burned slowly on rocks or in clay balls, and their ashes were then used as seasoning. Flavor and salt content can vary greatly with the habitat and age of a plant, so amounts added are usually based on personal taste and experimentation. Some strong-flavored plants are mixed with vinegar to make condiments, such as mustard and mint sauce.

Medicine

Plants have been used to cure, or at least to relieve the symptoms of, human diseases and injuries for 1000s of years. Ancient medicinal uses were often based more on superstition and folklore than on fact. In Europe, the 'Doctrine of Signatures' decreed that the use of a plant was revealed by its form, so hairy plants were good for stimulating hair growth, and plants with kidney-shaped leaves were used to treat kidney problems. Many of the uses recorded in ancient herbals were probably short lived. When one medicine failed, another was tried, but the first may still have been recorded as a treatment. In some parts of North America, it seems as though every common plant has been used to treat measles, smallpox, tuberculosis or syphilis. The historical use of a plant does not imply that it was effective. It could just have been one in a long series of failed experiments.

Many plants have persisted as medicines through years of trial and error because they proved effective. Some of these may have strengthened the patient by supplying important minerals and vitamins, but some contained powerful drugs that acted alone or in combination with other compounds. Modern medicine approaches drug plants from a more objective point of view, demanding standardized tests on the treatment of specific symptoms (rather than general, personal anecdotes) and analyzing the structure and action of the chemicals involved (rather than basing use on the knowledge passed down from earlier generations). This 'scientific' approach has discredited many 'natural' medicines, but there are also cases where research supports traditional uses. Most plants have yet to be studied.

Grindstone utensil for pounding berries

Preparations for herbal medicines are not described in detail in this book. Medicinal information is presented for interest's sake, rather than as a 'how-to' guide for a home pharmacy. Some plants can be used safely for treating minor injuries and illnesses, but many medicinal plants are poisonous if prepared improperly or taken in large doses. A person who is truly ill should consult a health-care professional, rather than experimenting with possible cures at home. Also, if you are taking medication, always check with your doctor before trying herbal remedies to avoid dangerous chemical interactions. For example, the drugs in the herbs could counteract the drugs in your medication, they could supplement your dose and raise it to dangerously high levels, or they could cause complications by creating antagonistic side effects. It takes years of study to understand the actions and interactions of medicinal herbs before putting this knowledge to use. If you are interested in learning more about the uses of medicinal plants, there are many books on this subject (see p. 259).

Plants are prepared and administered in many different ways as medicines. Drugs to be taken internally are usually prepared as medicinal teas. Water extracts are prepared by pouring boiling water over selected plant parts and then either leaving the mixture to steep for a short time to make an 'infusion,' or boiling it to produce a 'decoction.' When the active ingredient is not water-soluble, it is usually extracted by soaking the plant parts in alcohol (usually ethanol or vodka) for several hours or days to produce a 'tincture.' Many plant extracts are applied externally to combat infection and to treat a wide range of skin diseases, injuries, aches and pains. Some medicinal teas are used as washes to clean and heal injuries. Many plants (fresh or dried) are crushed, moistened and applied to injuries as hot or cold compresses or 'poultices.'

The concentration of active chemicals in a plant can vary greatly with its habitat and age. Some compounds are most concentrated in young shoots, while others increase as the plant ages and are most concentrated in mature plants. Many drugs are strongest in specific parts of the plant (e.g., in the roots, leaves or seeds) and are absent or much less concentrated elsewhere.

The way in which a plant material is prepared also affects the strength and effectiveness of the medicine produced. Many medicinal herbs are dried and stored for future use. Drying usually involves placing thin layers of plant parts on a well-ventilated surface (e.g., a screen or loosely woven fabric) in a warm, dry, dark place to ensure rapid drying and reduce chemical breakdown from exposure to light (photolysis). Thoroughly dried plants are then stored in airtight containers to reduce the loss of volatile compounds.

Plant material is often ground to produce herbal medicines. Traditionally, grinding involved crushing material between two rocks or with a mortar and pestle, but modern

Western blue flax

blenders, grain mills and coffee mills have greatly simplified the process. Some dried plants can be simply rolled into powder using a rolling pin.

Other Uses

Plants provide most of our foods and medicines, but they can also be used in many other ways. Plant fibers produce paper, clothing, thread and twine. Wood provides building materials and fuel. Light, airy plants and seed-fluff can provide insulation, bedding, tinder and absorbent padding for diapers and dressings. Even plant toxins can be used to our advantage for killing or repelling insects, rodents and other pests. Many plants are used for pleasure—chewed like gum, smoked like tobacco (*Nicotiana tabacum*) or used to make toy flutes and dolls for children. These applications are discussed under the heading 'Other Uses.'

Warning

Many plants have developed very effective protective mechanisms. Thorns and stinging hairs discourage animals from touching, let alone eating, many plants. Bitter, often poisonous compounds in leaves and roots repel grazing animals. Many protective devices are dangerous to humans. The Warning box includes notes of potential hazards associated with the plant(s) described. Hazards can range from deadly poisons to hairy leaves and airborne pollen that may cause allergic reactions. This section may also include descriptions of poisonous plants that could be confused with the species being discussed and notes about factors affecting toxicity (e.g., seeds or roots may be more toxic than other parts of the plant; young plants may be more dangerous than old plants).

The Poisonous Plants section (pp. 234–52) includes toxic species that have not been widely used by humans. Many plants discussed in the main body of the book are also poisonous, but they are included with the 'useful' species because they have been gathered as foods or medicines or used in other ways.

The fine line between delicious and dangerous is not always clearly defined. Many of the plants that we eat every day contain toxins. Broccoli, asparagus and spinach all contain carcinogenic alkaloids, but you would have to eat huge amounts in a short time to be poisoned. Almost any food is toxic if you eat enough of it. Personal sensitivities can also be important. People with allergies may die from eating common foods (e.g., peanuts) that are harmless to most of the population. Most wild plants are not widely used today, so their effects on a broad spectrum of society remain unknown.

As with many aspects of life, the best approach is 'moderation in all things.' Enjoy a varied diet with a mixture of different fruits, vegetables and grains. When trying something for the first time, take only a small amount to see how you like it and how your body reacts. Enjoy the wonderful array of foods that nature has provided, but sample it wisely!

Description

The plant description and the accompanying photos and illustrations are important parts of each species account. If you cannot correctly identify a plant, you cannot use it. Identification is more critical with some plants than with others. For example, most people recognize strawberries (pp. 128–29) and raspberries (p. 71) and all of the species in these two groups are edible (though not all are equally palatable). On the other hand, plants in the Carrot family (pp. 190–98) are much more difficult to distinguish, and some species contain deadly poisons, so confusion can have fatal results. Using a plant in this group, without positive identification, has been called 'herbal roulette.'

Each plant description begins with a general outline of the form of the species or genus named at the top of the page. Detailed information about diagnostic features of the leaves, flowers and fruits is then provided. Flowering time is also included as part of the flower description, to give some idea of when to look for blooms and, by extension, when fruits can be expected. If two or more species of the same genus have been used for similar purposes, two or three of the most common species in the region may be illustrated and their distinguishing features described.

As discussed earlier, plant characteristics such as size, shape, color and hairiness vary with season and habitat and with the genetic variability of each species. Identification can be especially tricky when plants have not yet flowered or fruited. If you are familiar with a species and know its leaves or roots at a glance, you may be able to identify it at any time

Figure I

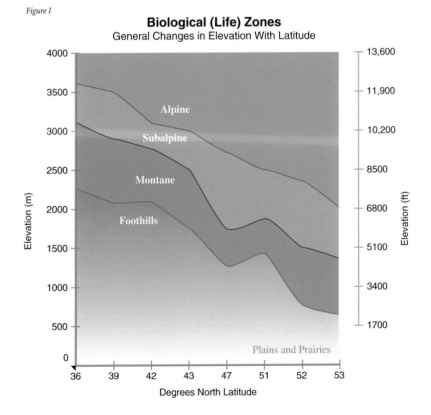

Biological (Life) Zones
General Changes in Elevation With Latitude

of year (from very young shoots to the dried remains of last year's plants), but sometimes a positive ID is just not possible.

General habitat information is provided for each species, to give you some idea of where to look for a plant. For example, if you are hiking through a subalpine fir forest, you probably won't find a plant whose habitat is described as 'dry slopes in the foothills.' However, the habitat information included for each species is meant as a general guide only. Mountain vegetation changes dramatically with elevation. Four main vegetation zones are referred to when describing plant distributions: foothills, montane, subalpine and alpine. In the mountains, climate and vegetation change dramatically with changes in elevation and/or latitude. A 186-mi (300-km) trip north has been equated with climbing up 328 ft (100 m). Consequently, the elevation of the major vegetation zones changes as you travel north or south through the Rockies (see Fig. I, p. 23). The vegetation of these zones is described in *Plants of the Rocky Mountains* (Kershaw et. al., 1998). Common plants often grow in a variety of habitats over a broad geographical range. Plant distribution is described for the Rocky Mountain region only. The abbreviations US, BC and NWT refer to the United States, British Columbia and the Northwest Territories respectively.

The origin of non–North American species is also noted. The flora of many areas in the Rocky Mountains has changed dramatically over the past 200 years, especially in and around human settlements. European settlers brought many plants with them, either accidentally (in ship ballast, packing and livestock bedding) or purposely (for food, medicine, fiber, etc.). Many of these introduced species have thrived in North America, and some are now common weeds on disturbed sites in the Rockies.

The
SPECIES

Firs

Abies spp.

Subalpine fir

FOOD: Cone fragments of subalpine fir were ground into powder and mixed with animal marrow or back fat. This mixture was cooled until it hardened and then served at social events as a delicacy that had the added benefit of aiding digestion. Inner bark of balsam fir (a mostly eastern species) was dried and ground into a nutritious (though not very tasty) meal and mixed with flour to extend food supplies in times of shortage.

MEDICINE: Dried, powdered needles were sprinkled on open, runny sores or mixed with deer grease to make salves for treating cuts, wounds, ulcers, sores, bleeding gums and skin infections. The needles are high in vitamin C and are also said to stimulate urination and to help to bring phlegm up from the lungs. Fir needle tea was taken for colds, and needles were used in poultices to treat fevers and chest colds. White fir branches were boiled to make a medicinal tea for stimulating urination and treating malaria. Smoke from burning subalpine fir needles was inhaled as a treatment for headaches, fainting and swollen faces (associated with venereal disease). Fir pitch was taken alone or in teas as a treatment for coughs, colds, asthma and tuberculosis. In stronger doses, it was said to induce vomiting and 'clean out the insides.' Fir pitch provides a handy antiseptic for wounds, cuts, boils, sores and bruises.

OTHER USES: Needle tea or dried, powdered needles were mixed with deer grease to produce a pleasant-smelling hair tonic. Powdered needles were also used as baby powder, or they were rubbed on the body and on clothing as perfume and insect repellent. Fir needles were burned as incense, and fir boughs were hung on walls as air-fresheners. Fir incense was thought to chase away bad spirits and ghosts and to revive the spirits of people near death. Burning fir wood was believed to give protection and renewed confidence to people who were afraid of thunder and lightning. The smoke was also said to cure sick horses. Subalpine fir resin provided a pleasant chew that cured bad breath. On grand fir, black conks (hoof-shaped, 'woody' fungal growths), called 'Indian paint fungus,' were used as a source of red pigment. Firs are a traditional Christmas tree in many areas, because they hold their needles well and they have a wonderful fragrance.

WARNING

Fir resin can cause skin reactions in some people. Always use evergreen teas in moderation. Do not eat the needles or drink the teas in high concentrations or with great frequency.

DESCRIPTION: Fragrant coniferous trees with whorled branches and thin, smooth young bark, with bulging resin blisters. Leaves flattened evergreen needles, spirally arranged but often twisted upwards or into 1 plane. Cones male or female, with both sexes on the same tree. Seed (female) cones cylindrical, stiffly upright, shedding scales with seeds and leaving a slender central core, appearing in May to July and maturing in 1 season.

Subalpine fir (*A. bifolia* or *A. lasiocarpa*) has thick, short (1–1¼" [2.5–3 cm]), sometimes pointed needles that curve upwards, giving the branch tips a rounded appearance. It grows on subalpine to alpine slopes from the southern Yukon to New Mexico.

Two other common firs have leaves twisted into 2 opposite rows, giving the branches a flattened, spray-like appearance.

White fir (*A. concolor*) has long (1⅝–2¾" [4–7 cm]), often pointed needles with white lines on both surfaces. It grows on montane slopes from southern Idaho and Wyoming to New Mexico.

Grand fir (*A. grandis*) has shorter (1¼–1⅝" [3–4 cm]), blunt to notched needles with white lines on the lower surface only. It grows in foothills and montane zones from southeastern BC to Idaho and Wyoming.

Subalpine fir (all images)

Picea spp.

FOOD: Spruce beer was popular among early northern travelers and was important in preventing scurvy. It is said to taste like root beer. Dried inner bark can be ground into a nutritious meal for extending flour during times of food shortage. It is best collected in springtime. Tender young shoots, stripped of their needles, can be boiled as an emergency food.

Seed cones of Engelmann spruce

MEDICINE: The sticky sap or moist inner bark was used in poultices on slivers, sores and inflammations. Similarly, sap mixed with fat provided salves for treating skin infections, insect bites, chapped hands, cuts, scrapes, burns and rashes. Melted sap was also used as plaster when setting bones. Spruce gum was chewed or was boiled and taken like cough syrup, to relieve coughs and sore throats. Some sources recommend spruce pitch as a sunscreen, but it seems like a rather sticky solution, not easily removed. Emerging needles were boiled to make an antiseptic wash or chewed to relieve coughs. Medicinal teas, made from the inner bark, were believed to cure rheumatism, kidney stones and stomach problems. Hot spruce needle tea is said to stimulate sweating, and the vapor was inhaled to relieve bronchitis.

OTHER USES: Spruce bark was sometimes used to make canoes. Lumps of hardened pitch were chewed for pleasure and to keep teeth white. Melted pitch was used to caulk canoe seams and to stick sheets of birch-bark or strands of willow-bark twine together. It was also used to preserve and waterproof strips of hide (babiche) in ropes, snares and snowshoes. Fresh or soaked spruce roots were peeled and split to make cord for stitching canoes and baskets and for making fishnets. Split roots were also woven into watertight bags for cooking. Spruce wood is generally strong and uniform. It is used for lumber, plywood, mine timbers, poles and railway ties. Its long fibers, light color and low resin content make it excellent pulp wood (black spruce is best). Because of its resonance, Engelmann spruce wood is used to make piano-sounding boards and violins. Some tribes used spruce trunks for tipi poles. Colorado blue spruce is a common ornamental.

Engelmann spruce

> **WARNING**
> Always use evergreen teas in moderation. Do not eat the needles or drink the teas in high concentrations or with great frequency. Some people develop rashes from contact with spruce resin, sawdust or needles.

DESCRIPTION: Coniferous, evergreen trees with thin, scaly bark. Leaves pointed, 4-sided needles, $3/8$–$1^1/4$" (1–3 cm) long, borne on persistent stubs and spirally arranged on all sides of the branches. Cones male or female, with both sexes on the same tree. Seed (female) cones with thin, flexible scales, hanging, produced in May to July, open in autumn and fall intact.

Black spruce (*P. mariana*) is a northern species of central BC and Alberta to Alaska. Its young twigs have tiny, rusty hairs, and its small ($3/8$–$1^1/4$" [1–3 cm]) seed cones persist on the tree for several years.

Engelmann spruce (*P. engelmannii*) usually has minutely hairy young twigs, and the scales of its $1^1/4$–3" (3–8 cm) long female cones have jagged upper edges. It grows on cool, moist montane and subalpine slopes from BC and Alberta to New Mexico.

White spruce (*P. glauca*) has hairless young twigs, and the scales of its 1–2" (2.5–5 cm) long seed cones have smooth, broadly rounded edges. It grows on foothill, montane and subalpine slopes from Alaska to northern Montana. White spruce and Engelmann spruce often hybridize.

Seed cone of Colorado blue spruce

Colorado blue spruce (*P. pungens*) has hairless young twigs, and the scales of its $2^1/2$–4" (6–10 cm) long female cones have jagged upper edges. It grows along drainage-ways from Idaho and Wyoming to New Mexico.

Seed cones of white spruce

Douglas-fir
Pseudotsuga menziesii

FOOD: The soft inner bark was used for survival food, and the small, pitchy seeds were also eaten. Young twigs and needles can be used as a substitute for coffee or tea. This mixture is usually sweetened with sugar. Sometimes on hot, sunny days, when photosynthesis and root pressure are high and transpiration is slow, white crystals of sugar appear at the needle tips and over the branches. This rare treat was usually eaten as a sweet nibble, but when quantities warranted it was collected and used to sweeten other foods.

MEDICINE: Dried sap was chewed to relieve cold symptoms, and sticky buds were chewed to heal mouth sores. Liquid pitch and the soft inner bark can be used on cuts, boils, sores and other skin problems to aid healing and prevent infection. Tea made from the gum or inner bark was used to treat colds. Young Douglas-fir needles contain vitamin C and were used in teas to treat scurvy. The bark was boiled to make a laxative tea.

OTHER USES: Fragrant Douglas-fir boughs were often used for bedding. The pliable roots have been used to weave baskets. Rotted wood from old logs was burned slowly to smoke hides, and the bark was used in tanning. Douglas-fir bark was also boiled with nets to make them brown and less visible to fish. Strong, durable Douglas-fir wood provides excellent lumber, plywood, poles and railway ties. The resin was sometimes burned as a fumigant, and resinous branches provided torches. Some tribes used the needles to induce vomiting during religious ceremonies. Other tribes placed the cones close to their fires to stop rain and bring sunshine. These fragrant trees are popular Christmas trees.

DESCRIPTION: Coniferous, evergreen tree with spreading or drooping branches. Mature trees with very open crowns. Buds pointed, shiny and reddish-brown. Young bark thin, smooth and resin-blistered. Older bark ridged and fissured. Leaves flat, blunt needles, $3/4$–$1^1/4$" (2–3 cm) long, spirally arranged, but often twisted into 2 rows. Cones male or female, with both on the same tree. Seed (female) cones 2–4" (5–10 cm) long, hanging, with 3-toothed bracts projecting beyond stiff scales, produced in April to May, opening in autumn and falling intact. Douglas-fir grows on foothill, montane and subalpine slopes from BC and Alberta to New Mexico.

WARNING
Always use evergreen teas in moderation. Do not eat the needles or drink the teas in high concentrations or with great frequency.

Western red cedar
Thuja plicata

FOOD: The moist inner bark (cambium) was collected in spring and eaten fresh or dried for future use.

MEDICINE: The bough tea, sweetened with honey, was used to relieve diarrhea, coughs, colds and sore throats. Tea made from the bark and twigs was used to treat kidney problems. The sprays are strongly anti-fungal and anti-bacterial. Alcoholic extracts are said to cure fungal skin infections such as athlete's foot, ringworm, jock itch and nail fungi.

Red-cedar teas and tinctures are reported to stimulate smooth muscles and vascular capillaries, so they are recommended by some herbalists for the treatment of problems associated with mucous accumulations or sluggishness in the respiratory, urinary, reproductive and digestive tracts. Steam from boiling sprays was sometimes inhaled by pregnant women to induce labor, and the green buds were chewed to relieve toothaches. Oil from the leaves has been applied to warts, hemorrhoids, fungal infections and *Herpes simplex* sores (watery blisters on the skin and mucous membranes).

OTHER USES: The bark peels off easily in long, fibrous strips. Native peoples twisted, wove and plaited the soft inner strips to make baskets, blankets, clothing, ropes, mats and other items. Sometimes the bark was boiled with wood ash to make the fibers softer, more easily separated and easier to work. Bark was also pounded until fluffy and used in diapers, sanitary pads and mattresses. Sheets of bark were formed into containers. Fine roots were split and peeled to make watertight baskets for cooking. Coiled larger roots were alternately bound together in 2s with split roots to make baskets. Dugout canoes, rafts and frames for birch-bark canoes were made from red cedar trunks. The light, easily worked wood splits readily and resists decay. It was made into cradle boards, bowls, masks, roofing, siding and many implements. Today it is widely used for siding, roofing, paneling, doors, patio furniture, chests and caskets. Cedar sprays make a lovely incense.

WARNING
Pregnant women and people with kidney disorders should not take red-cedar teas and extracts internally. Cedar oil can cause low blood pressure, convulsions and even death.

DESCRIPTION: Tall (to 130' [40 m]) coniferous tree with straight, gray, vertically lined trunk 3¹/₂–10' (1–3 m) in diameter. Leaves small (about ¹/₈" [3 mm]), overlapping, opposite scales forming, flat, yellowish-green, fan-like sprays. Cones male or female, with both sexes on the same tree. Seed (female) cones ³/₈–¹/₂" (8–12 mm) long, with 6–8 dry, brown scales, produced in April to May and maturing in autumn. Western red cedar grows in rich, moist to wet, foothill and montane sites in BC, Alberta, Idaho and Montana.

Western hemlock

Hemlocks

Tsuga spp.

FOOD: The sweet inner bark (cambium) was scraped from the trunk and baked or steamed in earth ovens. It was then pressed into cakes and eaten with fish oil or cranberries, or it was dried for future use. In winter, some tribes enjoyed the dried bark whipped together with snow and fish grease. Hemlock cambium can also be eaten raw in an emergency (though it is harder to digest) or dried and ground into meal to extend the amount of flour. The fresh needles make a fragrant evergreen tea, and small branch tips were sometimes cooked with bear meat as flavoring.

MEDICINE: Hemlock bark was boiled to make medicinal washes for treating burns and rashes. Teas made from the sweet inner bark were use to treat flu. The needles are rich in vitamin C, and they were used in teas to cure colds and relieve congestion. Hemlock oil was applied as liniment to relieve rheumatic pain.

OTHER USES: The soft, fragrant boughs were used for bedding and as a disinfectant and deodorizer. Boiled hemlock bark produced a preservative, red-brown dye for tanning hides, coloring wooden articles (e.g., spears, paddles) and making nets less visible to fish. The pitch was used to waterproof baskets and to protect skin from chapping and sunburn. The hard, strong wood, with its even grain and color, is used to make doors, windows, staircases, moldings, cupboards and hardwood floors. It also provides construction lumber, pilings, poles, railway ties, pulp and alpha cellulose for making paper, cellophane, rayon and some plastics. This attractive, feathery tree is a popular ornamental. Several cultivars have been developed.

> **WARNING**
> Always use evergreen teas in moderation. Do not eat the needles or drink the teas in high concentrations or with great frequency.

DESCRIPTION: Graceful coniferous, ever-green trees with feathery, down-swept branches and flexible, nodding crown-tips. Leaves small needles, about $^1/_4$–$^3/_4$" (1–2 cm) long, unequal, borne on small stubs. Cones male or female, with both sexes on the same tree. Seed (female) cones with thin, brownish scales, hanging, produced in May to June, opening in autumn and shed intact.

Western hemlock (*T. heterophylla*) has small, $^5/_8$–1" (1.5–2.5 cm) long seed cones and flat needles with white lines (rows of stomata) on the lower surface only. It grows on moist foothill and montane slopes in BC, Alberta, Idaho and Montana.

Mountain hemlock (*T. mertensiana*) has larger (1–2$^1/_2$" [2.5–6 cm]) seed cones and rounder needles with white lines on both sides. It grows on subalpine slopes in BC, Idaho and Montana.

Western hemlock (all images)

Larches

Larix spp.

Western larch

FOOD: Some tribes hollowed out cavities in western larch trunks to collect the trees' sweet sap. About 7 pt (4 *l*) could be taken from a good tree, once or twice a year. This sap was evaporated to the consistency of molasses; more recently, it was mixed with sugar to make syrup. The sweet inner bark was eaten in spring, and sweet lumps of dried sap were chewed like gum year-round. The gum is said to taste like candy. Larch sap contains galatan, a natural sugar with a flavor like slightly bitter honey. Dried, powdered larch gum was also used as baking powder. Tender young larch shoots can be cooked as a vegetable, and the inner bark can be dried, ground into meal and used to extend flour.

MEDICINE: Larch gum was chewed to aid digestion and to relieve sore throats, internal bleeding and enlarged, hardened livers. The gum and/or the soft inner bark was used in poultices to treat insect bites, cuts, bruises, wounds, ulcers and persistent skin problems such as eczema and psoriasis. It was also applied externally and taken internally in teas to relieve rheumatism. Larch bark tea was taken to treat jaundice, colds, coughs, bronchitis, tuberculosis and asthma, and tea made from the bark and needles was said to cure both diarrhea and constipation.

OTHER USES: Rotted larch logs were often a preferred source of fuel for smoking hides, because the finished buckskins were neither too dark nor too light. Some tribes chose the larch for their center pole in the Sundance, whereas tribes further east chose the cottonwood. Today, the straight, relatively branch-free trunks, with their heavy, durable wood, are often used for telephone poles, railway ties, posts, construction timbers and mine timbers. Because larch wood was reputed not to rot, it was often used in boat-building. Arabino galatin, a water-soluble gum, is extracted from larch bark for use in paint, ink and medicines

WARNING
Larch resin and sawdust cause skin reactions in some people. Always use evergreen teas in moderation. Do not eat the needles or drink the teas in high concentrations or with great frequency. Some tribes warned that eating too much of the sweet inner bark would 'clean you out.'

DESCRIPTION: Slender, coniferous trees with short, well-spaced, sparsely leaved branches. Leaves soft, deciduous needles, 1–1³/₄" (2.5–4.5 cm) long, in tufts on stubby twigs, bright yellow in autumn and shed for winter. Cones male or female, but both sexes on the same tree. Seed (female) cones with slender, 3-pointed bracts projecting beyond thin, woody scales, red when young, produced in May to July, mature by autumn.

Western larch (*L. occidentalis*) has essentially hairless young twigs and small (1–1¹/₄" [2.5–3 cm]) seed cones, and its needles are flat-topped but strongly ridged beneath. It grows on well-drained upper foothills and montane slopes from BC and Alberta to Oregon and Montana.

Subalpine larch (*L. lyallii*) has wooly young twigs, larger (1³/₈–1³/₄" [3.5–4.5 cm]) cones and 4-sided needles. It usually grows at higher elevations (near timberline) over a similar range.

Western larch (all images)

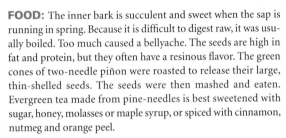

Two-needled pines

Pinus spp.

FOOD: The inner bark is succulent and sweet when the sap is running in spring. Because it is difficult to digest raw, it was usually boiled. Too much caused a bellyache. The seeds are high in fat and protein, but they often have a resinous flavor. The green cones of two-needle piñon were roasted to release their large, thin-shelled seeds. The seeds were then mashed and eaten. Evergreen tea made from pine-needles is best sweetened with sugar, honey, molasses or maple syrup, or spiced with cinnamon, nutmeg and orange peel.

MEDICINE: Pine-needle tea is high in vitamins A and C and was taken in winter to prevent or cure scurvy. Some tribes used strong pine-needle tea as a contraceptive. The inner bark was used as a dressing for scalds, burns and skin infections. Pine pitch was chewed to soothe sore throats and sweeten bad breath and was taken internally to treat kidney problems and tuberculosis. Warmed sap was applied to sore muscles, arthritic joints, swellings and skin infections. It was also heated until it turned black, mixed with bone marrow (1 part marrow to 4 parts sap) and used as a salve for burns. A poultice of pine sap, red axle grease and Climax chewing tobacco (*Nicotiana tabacum*) was recommended for boils!

OTHER USES: The straight, slender trunks of lodgepole pine were used as poles for travois, tipis and lodges (hence the name 'lodgepole pine'). Each tipi required 25–30 poles, 23–26' (7–8 m) long. These usually had to be replaced almost every year and tribes from the plains sometimes traveled hundreds of miles to the mountains to get new poles. Lodgepole pine wood is light colored and straight grained, with a soft, even texture. Today, pine lumber is used to frame homes and construct moldings and furniture.

DESCRIPTION: Coniferous trees with $^3/_4$–$2^1/_2$" (2–6 cm) long, evergreen needles in bundles of 2. Seed cones thick-scaled, oval, mature in 2 years, but may remain closed on branches for many years.

Lodgepole pine (*P. contorta*) has $1^1/_4$–$2^1/_2$" (3–6 cm) long cones with spine-tipped scales. It is widespread in foothills and montane zones from Alaska to Colorado.

Two-needle piñon (*P. edulis*) has 1–2" (2.5–5 cm) long cones with spineless scales. It is common in dry regions from Utah south.

> **WARNING**
> Large amounts of evergreen tea can be toxic. Pregnant women should not drink this tea.

Lodgepole pine (all images)

Ponderosa pine

Pinus ponderosa

FOOD: The sweet inner bark (cambium) was said to taste something like sheep fat. It was collected on cool, cloudy days when the sap was running. The bark was removed from one side of a tree only (to avoid killing the tree), and the edible inner bark was then scraped from the tough outer layer. This inner bark was usually eaten immediately, but it could also be kept moist, rolled in bags, for a few days. Some pines still bear large scars—tell-tale signs of early native peoples' fondness for pine cambium. The oil-rich seeds are also edible. They were shaken from the cones and ground into meal that was used to make bread. The young, un-opened male cones can be boiled as an emergency food. The young needles, finely chopped, can be used to make tea. Some say that ponderosa pine makes the best pine tea, and others say that it is the most potentially toxic.

MEDICINE: The resin was applied alone or in salves to boils, carbuncles, abscesses, rheumatic joints and aching backs. For dandruff, the pointed ends of the needles were jabbed into the scalp to kill the germs. The needle tea is rich in vitamins A and C.

OTHER USES: The pitch was chewed as gum, plastered in hair, used as glue, burned in torches and used to water-proof woven containers. The light, soft wood has been used to make dwellings, fence posts, saddle horns, snowshoes and baby cradles. Today, it is used as lumber and made into moldings, cabinets and crates.

DESCRIPTION: Coniferous tree with evergreen needles 4–10" (10–25 cm) long, usually in bundles of 3. Bark orange-brown to cinnamon with jigsaw-like plates outlined by deep, black fissures. Seed cones oval, 8–14 cm long, with thick, spine-tipped scales, maturing in 2 years. Ponderosa pine grows on dry sites in foothills and montane zones from southern BC to New Mexico.

WARNING
Large amounts of pine tea can be toxic, and extended use irritates the kidneys. Pregnant cows that eat ponderosa pine needles may abort their calves in 2 days to 2 weeks. Pregnant women should not drink this tea.

Five-needled pines

Pinus spp.

FOOD: The inner bark (cambium) is sweet and edible in spring, and young needles can be steeped to make pine tea. The oil-rich, nutritious seeds were eaten by native peoples and early settlers, but the cones of whitebark pine were generally considered too small and too greasy to warrant the effort necessary. Unopened cones were burned in a large fire, which cooked and released the seeds at the same time. The seeds were then eaten, ground into meal for making breads or stored for later use.

MEDICINE: The tender young needles are a good source of vitamin C. The Navajo used pine-needle tea to treat fevers and coughs. This tea is said to stimulate urination and to aid in bringing up phlegm and mucous from the lungs. Pine resin has been used for many years in cough syrups and in ointments for burns and skin infections.

OTHER USES: Several cultivars of limber pine, including a dwarf form, are used as ornamental trees. The Navajo burned limber pine to produce good-luck smoke for hunters.

DESCRIPTION: Coniferous, often gnarled, twisted trees with evergreen needles 1⅝–2¾" (4–7 cm) long, in bundles of 5. Seed cones thick-scaled, containing large (⅜" [1 cm] long) seeds.

Immature cones (bottom photo)

Limber pine (*P. flexilis*) has light brown, 3–10" (8–25 cm) long cones that shed their seeds and fall intact. It grows in exposed foothill to subalpine sites from southern BC and Alberta to New Mexico.

Whitebark pine (*P. albicaulis*) has purplish-brown, 2–3" (5–8 cm) long cones that usually remain closed and gradually break apart. It grows on exposed slopes near timberline from southern BC and Alberta to Wyoming.

WARNING
Large amounts of evergreen tea can be toxic. Pregnant women should not drink this tea. Extended use irritates the kidneys.

Limber pine (all images)

White birch

Betula papyrifera

ALSO CALLED: paper birch.

FOOD: These trees produce large amounts of sap in spring that can be used straight from the tree as a beverage. The sap has also been boiled to make syrup, although it contains only about half as much sugar as maple sap. Thin syrup or the sap mixed with sugar or honey was fermented to make birch beer, wine or vinegar. Young twigs and bark were boiled to make tea. The tea was also sweetened with sugar or honey and made into beer or vinegar. The sweet inner bark was added to soups and stews or was dried and ground into flour for making bark bread. Young leaves and catkins were sometimes used to flavor salads, meat dishes and cooked vegetables.

MEDICINE: Recent research indicates that betulic acid (the compound that makes birch-bark white) may be useful for treating skin cancer. It could become an ingredient in future sunscreens and tanning lotions. The leaves, twigs and buds contain salicylates, so they have been used to make medicinal teas for relieving pain and inflammation. The leaf tea is also said to stimulate urination.

OTHER USES: The thin, paper-like bark of these trees was widely used by native peoples to make canoes, baskets, storage containers, cups and platters. Heating makes it amazingly pliable. Thin strips of birch-bark make excellent kindling, and the wood produces long-burning, sweet-smelling fires. Birch wood is extremely hard. It was traditionally used to make sleds, snowshoes, paddles, canoe ribs, arrows, tool handles and even needles. Birch trunks were sometimes used as poles for making tipis and drying racks. Do not cut birch-bark from living trees; bark removal can permanently scar and even kill birches.

DESCRIPTION: Small deciduous tree with smooth, white to yellowish bark that peels off in papery sheets. Leaves $1^5/8$–$3^1/2$" (4–9 cm) long, yellow in autumn. Flowers tiny, in dense, slender clusters (catkins), with male and female catkins on the same tree. Pollen (male) catkins 2–4" (5–10 cm) long, loosely hanging. Seed (female) catkins $3/4$–$1^1/2$" (2–4 cm) long, erect, shedding winged nutlets and 3-lobed scales, in April to May. White birch grows on dry to moist sites in foothills to subalpine zones from Alaska to Colorado.

Balsam poplar

Populus balsamifera

ALSO CALLED: black cottonwood • balm-of-Gilead • *P. trichocarpa.*

FOOD: Many tribes relished the sweet inner bark (cambium) of balsam poplar in spring, when the sap was running. After the thick outer bark had been removed, a buffalo or elk rib was used to scrape off the translucent inner bark. Hollows were sometimes carved in trunks to collect the sap. The young catkins were also eaten.

MEDICINE: The leaves were applied to bruises, sores, boils and aching muscles and to sores on horses that had become infected with maggots. The bark was chewed to relieve colds or was used in teas for treating tuberculosis and whooping cough. Some tribes boiled the bark tea to make a thick syrup that was spread on cloth to make casts for supporting broken bones. Bark ashes, mixed with water and cornmeal, provided a poultice for boils. Resins from the sticky, aromatic buds (gathered in late winter and spring) have been used for centuries in salves, cough medicines and pain killers. Salves, made by mixing the buds with fat, were used to relieve congestion from colds and bronchitis, to heal skin infections and to soothe aching muscles.

OTHER USES: Balsam poplar wood was said to be ideal for tipi fires, because it did not crackle and it made clean smoke. However, poplar trunks are often punky from fungal infections. In spring, buds mixed with blood produced a permanent black ink that was used to paint records on traditional hide robes. Twigs and bark were occasionally fed to horses when forage was limited. When stealing horses, war parties carried balsam poplar bark with them to feed the horses. Some warriors rubbed themselves with the sap to mask their scent. In some regions, these fast-growing trees are now grown as a crop for pulp.

DESCRIPTION: Deciduous tree with deeply furrowed mature bark and large, resinous, fragrant buds. Leaves 2–4 3/4" (5–12 cm) long, round-stalked, dark green above and paler beneath. Flowers tiny, in hanging clusters (catkins), with 3/4–1 1/4" (2–3 cm) long male catkins and 1 5/8–4" (4–10 cm) long female catkins on separate trees, appearing in April to May and producing oval capsules that release fluffy masses of tiny seeds tipped with soft, white hairs. Balsam poplar grows on moist to wet sites, often on shores, in foothills to subalpine zones from Alaska to Colorado.

WARNING
Some sources report that the bark tea may be slightly toxic. The bud resin may irritate sensitive skin.

Trembling aspen

Populus tremuloides

ALSO CALLED: aspen poplar.

FOOD: In spring, when the sap began to flow, the pulpy inner bark (cambium) provided a sweet treat for children of some northern tribes. It was scraped off in long strips and eaten raw. The bitter leaf buds and young catkins are edible and are rich in vitamin C.

MEDICINE: Bark tea has been used to treat skin problems, intermittent fevers, urinary tract infections, jaundice, debility and diarrhea and to kill parasitic worms. Some tribes believed that the bark had to be collected by stripping it downwards from the tree; otherwise the patient would vomit it out! The leaves and inner bark contain salicin, a compound similar to salicylic acid, so these parts have been used, like aspirin, to relieve pain, fever and inflammation. Syrup made from the inner bark was taken as a spring tonic and as a cough medicine.

OTHER USES: Aspen trunks were used occasionally as tipi poles. The light, odorless wood shreds and splits easily, so it has been used to make paddles, bowls, wooden matches, vegetable crates, wallboard and excelsior for packing. Today, aspen wood is primarily harvested for pulp and for making chopsticks. Because it does not splinter, it is preferred for sauna benches and playground equipment. Aspen is an attractive, fast-growing tree, but it is seldom used in landscaping because it is susceptible to many diseases and insects, and its spreading root system tends to grow where it is not wanted (invading sewers and drainpipes). In spring, when the bark can be slid from the branches, short sections can be used to make toy whistles.

DESCRIPTION: Slender deciduous tree with smooth, greenish-white bark, marked with blackened spots and lines. Buds small, not resinous. Leaves ³/₄–3" (2–8 cm) long, with slender, flattened stalks that cause them to tremble in the slightest breeze. Flowers tiny, in hanging, ³/₄–4" (2–10 cm) long clusters (catkins), with male and female catkins on separate trees, appearing in March to May, producing cone-shaped capsules that release many tiny seeds tipped with soft, white hairs. Aspen grows in dry to moist sites in foothills to subalpine zones from Alaska to New Mexico.

Common juniper

Juniperus communis

FOOD: Juniper berries can be quite sweet by the end of their second summer or in the following spring, but they have a rather strong, 'pitchy flavor' that some people find distasteful. They are usually added as flavoring in meat dishes (recommended for venison and other wild game, veal and lamb), soups and stews, either whole, crushed or ground and used like pepper. For more information, see cedar-like junipers (pp. 44–45).

MEDICINE: Oil-of-juniper from juniper berries was mixed with fat to make salves for protecting wounds from irritation by flies. Juniper berries stimulate urination by irritating the kidneys and give the urine a violet-like fragrance. They are also said to stimulate sweating, mucous secretion, production of hydrochloric acid in the stomach and contractions in the uterus and intestines. Some studies have shown juniper berries to lower blood sugar caused by adrenaline hyperglycemia, suggesting that they may be useful in the treatment of insulin-dependent diabetes. Juniper berries have antiseptic qualities, and studies by the National Cancer Institute have shown that some junipers contain antibiotic compounds that are active against tumors. Strong juniper tea was used to sterilize needles and bandages, and during the Black Death in 14th-century Europe, doctors held a few berries in the mouth to avoid infection from patients. During cholera epidemics in North America, some people drank and bathed in juniper tea to prevent infection. Juniper tea was often given to women in labor to speed delivery, and after the birth it was used as a cleansing, healing agent. Juniper needles were dried and powdered as a dusting for skin diseases, and juniper smoke or steam was inhaled to relieve colds and chest infections. See Other Uses under cedar-like junipers.

OTHER USES: Juniper branches were often burned in smudges to repel insects. Smoke from the berries or branches of junipers was used in religious ceremonies or to bring good luck (especially for hunters) or protection from disease, evil spirits, witches, thunder, lightning and so on. If a horse

WARNING:
Large and/or frequent doses of juniper can result in convulsions, kidney failure and an irritated digestive tract. People with kidney problems and pregnant women should never take any part of a juniper internally. Juniper oil can cause blistering.

was sick, it was made to inhale smoke from juniper needles 3 times for a cure. The berries make a pleasant, aromatic addition to potpourris, and vapors from boiling juniper berries in water were used like the smoke to purify and deodorize homes affected by sickness or death. Juniper boughs were sometimes hung in tipis for protection from thunder and lightening. See Other Uses under cedar-like junipers.

DESCRIPTION: Spreading, 1–3$^1/_2$' (30–100 cm) tall evergreen shrub with small, $^1/_4$–$^3/_8$" (5–9 mm), bluish-purple to bluish-green 'berries' (actually fleshy cones), produced in April to May and maturing the following year. Leaves sharp, $^1/_4$–$^1/_2$" (5–12 mm) long needles, in whorls of 3. Common juniper grows on dry, open sites in plains to alpine zones from Alaska to New Mexico.

Cedar-like junipers
Juniperus spp.

FOOD: Some tribes cooked juniper berries into a mush and dried it in cakes for winter use. The berries were also dried whole and ground into meal for making mush and cakes. Dried, roasted juniper berries were ground and used as a coffee substitute. Teas were occasionally made from the stems, leaves and/or berries, but they were usually used as medicines rather than beverages. Juniper berries are best known for their use as a flavoring for gin, beer and other alcoholic drinks. Tricky Marys can be made by soaking juniper berries in tomato juice for a few days and then following your usual recipe for Bloody Marys, but omitting the gin. The taste is identical and the drink is non-alcoholic. For more information, see common juniper (pp. 42–43).

MEDICINE: Juniper-berry tea has been used to aid digestion, to stimulate appetite, to relieve colic and water retention, to treat diarrhea and heart, lung and kidney problems, to prevent pregnancy, to stop bleeding, to reduce swelling and inflammation and to calm hyperactivity. Juniper berries were chewed to relieve cold symptoms, settle upset stomachs and increase appetite. On the other hand, small pieces of the bitter bark or a few berries were chewed in times of famine to suppress hunger. Tea made from the branches and cones was used to treat fevers, colds, coughs and pneumonia. This tea was also heated to soak arthritic and rheumatic joints, or it was mixed with a bit of turpentine and applied as a liniment. Oil-of-juniper, diluted with less-irritating oils, was often used in liniments. It is still found in some patent medicines. For more information, see common juniper.

Creeping juniper (top)
Rocky Mountain juniper (above)

OTHER USES: Juniper berries were sometimes dried on strings, smoked over a greasy fire and polished to make shiny black beads for necklaces. Some tribes scattered berries to be used for necklaces on anthills. The ants would eat out the sweet center, leaving a convenient hole for stringing. The durable wood of Rocky Mountain juniper was used in lance shafts, bows and other items. It was admired for its dark red, seemingly-dyed-in-blood color. Today, this

WARNING:
Large and/or frequent doses of juniper can result in convulsions, kidney failure and an irritated digestive tract. People with kidney problems and pregnant women should never take any part of a juniper internally. Juniper oil can cause blistering.

attractive shrub is often grown as an ornamental. Juniper branches were sometimes boiled to produce an anti-dandruff hair rinse. The bark, berries and needles produced a brown dye, using ash from burned green needles as a mordant. For more uses, see common juniper.

DESCRIPTION: Coniferous evergreens with opposite rows of small ($^1/_8$" [1–1.5 mm] long), scale-like leaves. Fruits small, $^1/_4$–$^3/_8$" (5–9 mm), bluish-purple to bluish-green 'berries' (fleshy cones), produced in May to June and maturing the following year.

Rocky Mountain juniper

Rocky Mountain juniper (*J. scopulorum*) is an erect shrub or small tree, $3^1/_2$–33' (1–10 m) tall. Its young leaves are often $^1/_4$" (5–7 mm) long and needle-like, but the mature leaves are tiny and scale-like. It grows on dry, open foothill and montane slopes from BC and Alberta to New Mexico.

Creeping juniper or **ground juniper** (*J. horizontalis*) is a low shrub (seldom over 6" [15 cm] tall) with trailing branches, bearing tiny, scale-like leaves in 4 vertical rows. It grows on dry, open slopes in plains to subalpine zones from the Yukon and NWT to Wyoming.

Rocky Mountain juniper

Creeping juniper

Western yew
Taxus brevifolia

FOOD: The fleshy red parts of these fruits is said to be edible, but the seeds are extremely poisonous, so this fruit is not recommended (see Warning).

MEDICINE: The bark of western yew is a source of taxol (paclitaxel), an alkaloid that was discovered to inhibit the growth of some cancers in mice. It has been studied as a potential drug for combating leukemia and solid tumors and has been given intravenously to treat cancer of the ovaries, cervix, skin (malignant melanomas) and breast. Some native peoples used yew bark for treating illness and applied the wet needles as poultices on wounds—do not try this remedy (see Warning).

OTHER USES: The heavy, fine-grained wood has been used to make bows, wedges, clubs, paddles, digging sticks, tool handles and harpoons. Two ancient spears, dating from the early Stone Age, were found to be made of yew wood. The wood is still prized today by carvers, but it is relatively scarce. Bows, carved from seasoned yew wood, were varnished with boiled animal sinew and muscle. With their dark, evergreen needles and scarlet berries, yews make lovely ornamental shrubs, but their poisonous seeds, branches and leaves could be dangerous to children.

DESCRIPTION: Evergreen shrub or small tree, 6¹/₂–33' (2–10 m) tall, with papery, reddish-purple bark and drooping branches. Leaves flat, ¹/₂–³/₄" (14–18 mm) long needles, twisted into 2 opposite rows. Flowers tiny, ¹/₈" (2–3 mm) long, in male and female cones on separate shrubs, produced in April to June and mature in autumn. Fruits scarlet 'berries' (arils), ¹/₈" (4–5 mm) across, with a cup of fleshy tissue around a single, bony seed. Western yew grows in moist, shady sites in foothills and montane zones from BC and Alberta to Idaho and Montana.

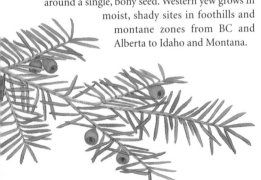

WARNING:
The leaves, bark and seeds contain taxine, a heart-depressing alkaloid that is extremely poisonous. Drinking yew tea or eating as few as 50 leaves can cause death. The berries are eaten by many birds, and the branches are said to be a preferred winter browse for moose, but many horses, cattle, sheep, goats, pigs and deer have been poisoned from eating yew shrubs, especially when the branches were previously cut.

Oaks

Quercus spp.

FOOD: Acorns were eaten raw, boiled, roasted in ashes or dried and ground into meal. White acorns, such as those of Gambel's oak, are sweetest, but even they were often too bitter to eat fresh from the tree. The outer shells were removed and the inner kernels (fresh or dried, crushed or whole) were placed in a mesh bag in water and soaked for several hours (or weeks). Sometimes wood ashes were added to speed the process, or kernels were boiled in several changes of water until the water no longer turned brown. Leaching removes the tannin, but most of the important food components remain. Some tribes added special clays to mask the bitterness. Powdered gelatin may also work. Acorn meal was a staple food of some tribes. It was used to make mush, breads, muffins and pancakes and to thicken soups. Roasted acorn kernels taste much like other nuts and can be eaten as a snack, baked in cookies and cakes, or dipped in syrup and eaten like candy. Dried kernels and meal keep for several months. Acorn shells and/or kernels were roasted, ground and brewed to make a coffee-like drink.

Gambel's oak

MEDICINE: Oak bark tea is high in tannins and quercin, and swellings on the twigs (insect-caused galls) contain 2–3 times as much of these compounds as regular bark. Oak bark teas have been widely used in washes and gargles for treating inflamed gums, sore throats, burns, cuts, scrapes, insect bites and rashes and in medicines and enemas for treating diarrhea, hemorrhoids and menstrual problems. Quercin is said to strengthen capillaries, and tannic acid is anti-viral and anti-bacterial. Pieces of oak bark were chewed to relieve toothache pain. Oak root bark was boiled to make medicinal teas for purging the system, for speeding delivery of afterbirth, for relieving pain after childbirth and for regulating bowel problems (especially in children).

OTHER USES: Oak wood was used to make handles for axes and hoes, digging sticks, weaving tools, bows and arrows, baby cradles and ceremonial bullroarers.

DESCRIPTION: Deciduous or evergreen trees or shrubs with alternate, usually pinnately lobed or toothed leaves. Flowers tiny, without petals, either male or female, but with both sexes on the same plant. Male (pollen) flowers in dense, elongated clusters (catkins) and female (seed) flowers single, clumped or in catkins. Fruits 1-seeded nuts (acorns), each seated in a thick, scaly cup. Oak species can be very difficult to identify.

WARNING:
Tannic acid is potentially poisonous. Cattle and sheep have died from eating oak leaves or large amounts of acorns.

Gambel's oak (*Q. gambelii*) was the most common and widely used species in the Rockies. It is a 10–16' (3–5 m) tall shrub or small tree, with deeply lobed, 2–4" (5–10 cm) long leaves and ¹/₂–⁵/₈" (12–15 mm) long acorns on which the cap covers ¹/₃ or more of the nut. It grows on dry foothill slopes from southern Wyoming to New Mexico.

Willows

Salix spp.

FOOD: The young shoots and leaves, buds and inner bark are edible, though rather bitter, and they can be used raw or cooked as an emergency food that is rich in vitamin C. The inner bark is said to be more palatable dried and ground into flour. In the Arctic, half-digested willow twigs from the stomachs of slaughtered caribou were considered a special treat.

MEDICINE: Willow bark has been used for centuries to relieve pain, inflammation and fever. It contains salicin, which is similar to acetylsalicylic acid (aspirin). Willow bark was chewed or made into medicinal teas for treating diarrhea and other digestive problems, headaches, arthritis, rheumatism and urinary tract irritations. Aspirin has been shown to delay the formation of cataracts

Arctic willow

and to reduce the risk of heart disease in men. Willow bark is less potent than aspirin but also has fewer side effects (eg., less stomach upset, no impact on blood platelet function). Some herbalists recommend taking willow bark to lose weight, but there is no evidence to support this use. Bark tea or strips of bark softened by chewing or boiling were used in washes and poultices on minor burns, insect bites, cuts, scrapes, rashes, ulcers, corns and even cancers. In Ancient Rome, ash from willow leaves or twigs was believed to remove corns, calluses and facial blemishes, and mashed leaves were taken with drink to check overly lustful behavior (though overdoses could cause impotence).

OTHER USES: Flexible willow branches are easily found. They were used to make many common articles, including pins, pegs, backrests, mattresses, fish traps, fox traps, cradle boards, snowshoes, gambling wheels, walking sticks, stirrups, hide-scrapers, hoops for catching horses, baskets, drums, meat racks and frames for sweat houses and temporary tipis. Willow bark was stripped and twisted to make twine, baskets and fishing nets that were strong when wet, but brittle when dry. Willow bark was chewed to clean teeth and prevent cavities. Dried, crushed willow roots were soaked in water and mixed with grease to make a dandruff tonic. Leaves and twigs were boiled to make a hair rinse for curing dandruff, or the hair rinse was sometimes mixed with wine and used as shampoo. Willow whistles can be made in spring. Simply cut a short branch section and make a notch near one end. Then slip off the bark, remove a thin slice of wood between the notch and the tip of the wood, replace the bark and blow through the hole at the end. The prolific male catkins offer an early spring feast for bees.

WARNING: People who are sensitive to aspirin should not take willow internally. Large amounts irritate the stomach and contain tannin, which is potentially cancer-causing in high doses.

DESCRIPTION: Deciduous shrubs with a single scale covering each bud. Flowers tiny, in dense, fuzzy clusters (catkins, pussy willows), with male or female catkins on separate shrubs, produced in spring or early summer and mature in a few weeks. Fruits small capsules in dense catkins, releasing many tiny seeds, each tipped with a tuft of silky hairs. This large, highly variable group of shrubs has dozens of species in the Rockies.

The tiny **Arctic willow** (*S. arctica*), rarely more than 4" (10 cm) tall, grows on subalpine and alpine slopes from Alaska to New Mexico.

Catkins of Bebb's willow

Of the larger willows, **Scouler's willow** (*S. scouleriana*) can be 30' (9 m) tall. It has broad-tipped leaves with short, stiff, reddish hairs on the lower surface.

Sandbar willow (*S. exigua*) is a flood-plain shrub with linear leaves 2–5" (5–13 cm) long.

Bebb's willow (*S. bebbiana*) has long-stalked capsules in loose catkins, and its young leaves have netted, raised veins on the lower surface.

All 3 of these willows grow on montane slopes and at lower elevations from Alaska to New Mexico.

Scouler's willow

Sandbar willow

Alders

Alnus spp.

FOOD: Although the buds and young catkins are not very tasty, they are edible and high in protein and can be eaten in an emergency. Similarly, the inner bark may be eaten raw, cooked or dried and ground into flour.

MEDICINE: Alder bark tea was taken hot to treat tuberculosis of lymph nodes (scrofula) and to regulate menstrual periods and relieve cramping. It was also used as a wash or douche for healing hemorrhoids and vaginal infections. Tea made by boiling the leaves or inner bark has been used as a soothing wash for insect bites, poison-ivy rashes, sores, burns and other skin problems. Bags of heated alder leaves have been used as hot compresses on people suffering from rheumatism. Some tribes chewed the bark and used it as a poultice on cuts, wounds and swellings. The ancient Romans recommended alder leaves for treating tumors. Scientists have discovered 2 tumor-suppressing compounds, betulin and lupeol, in red alder (*A. rubra*), a close western relative.

OTHER USES: Alder was a preferred fuel for smoking fish, meat and hides. It makes excellent fires, with few sparks and little ash. The twigs and inner bark were used to dye hides, feathers, moccasins and fishnets reddish-brown or orange. Dyed nets were thought to be harder for fish to see. Boiling produced more brilliant colors, and bark tannin set the color, so no mordant was needed. Some people even dyed their hair red with alder bark. The bark was sometimes chewed, spat onto hides and rubbed in for coloring. Its high tannin content also helped the tanning process. Alder leaves were used as insoles, to ease tired feet. Young boys enjoyed chewing alder buds in spring and spitting jets of dark juice onto the snow—just like real chewing tobacco (*Nicotiana tabacum*). If

Green alder (both photos)

WARNING:
Green bark can cause vomiting and sharp pains in the bowels. Bark should be aged for at least a few days prior to use, and bark decoctions were often left to sit until they have changed from yellow to black (2–3 days).

someone wanted to appear to be spitting blood, he or she could chew the inner bark and spit red saliva. Moist alder leaves are said to attract fleas, which can then be killed or at least moved outside.

DESCRIPTION: Clumped, deciduous shrubs or small trees with alternate, sharply toothed leaves $1^5/_8$–4" (4–10 cm) long. Flowers tiny, in dense clusters (catkins), with male and female catkins on the same branch. Pollen (male) catkins, slender and loosely hanging. Seed (female) catkins woody, cone-like, $^1/_2$–$^3/_4$" (1–2 cm) long, produced in May to July. Fruits tiny nutlets in the seed cones.

Green alder (*A. viridis*) has dark yellowish-green, often shiny leaves with single or double teeth, and its seed cones develop with the leaves on new twigs. It grows in moist foothill, montane and subalpine sites from Alaska to Colorado.

Mountain alder (*A. incana*) has dull green, wavy-lobed, double-toothed leaves, and its seed cones appear before the leaves on twigs from the previous year. It grows near water in foothills, montane and subalpine zones from Alaska to New Mexico.

Green alder

Mountain alder

Flowers of northern gooseberry

Gooseberries

Ribes spp.

FOOD: All gooseberries and currants (*Ribes* spp.) are edible raw, cooked or dried, but flavor and sweetness vary greatly with species, habitat and season. All are high in pectin and make excellent jams and jellies. Gooseberries are often eaten fresh, and they are delicious alone or mixed with other fruit in pies. Although gooseberries are common and widespread, they were not usually collected in large quantities by native peoples. Dried gooseberries were sometimes included in pemmican. Occasionally, gooseberries were mashed (usually in a mixture with other berries) and formed into cakes that were dried and stored for winter use. Dried gooseberry-and-bitterroot (p. 118) cakes were sometimes traded with other tribes. Because of their tart flavor, gooseberries can be used much like cranberries. They make a delicious addition to turkey stuffing, muffins and breads. Timing is important when picking gooseberries. Green berries are too sour to eat and ripe fruit soon drops from the branch. Sometimes green berries are collected and left to ripen. For more information, see currants (pp. 54–57).

MEDICINE: Gooseberries and currants are rich in vitamin C. They were commonly eaten or used in teas for treating colds and sore throats. Teas made from gooseberry leaves and fruits were given to women whose uteruses had slipped out of place after too many pregnancies. Gooseberry tea was also used as a wash for soothing skin irritations such as poison-ivy rashes and erysipelas (a condition with localized inflammation and fever caused by a Streptococcus infection). Both currants and gooseberries have strong antiseptic properties. Extracts have proved effective against yeast (Candida) infections, and some show promise in cancer treatment. Russian studies are reported to have shown that unripe gooseberries can prevent body cells from degenerating—could these berries be the end of aging and disease?

OTHER USES: Gooseberry thorns were often used as needles for probing boils, removing splinters and applying tattoos. In the early 1900s, 10 million pine seedlings infected with blister rust (*Cronartium ribicola*) were brought to North America from Europe, and the rust soon spread to infect the native, 5-needled pines. To reproduce, rust must spend part of its life cycle on an alternate host. *Ribes* species are the alternate host for blister rust, so a program to eradicate these shrubs was attempted, unsuccessfully.

WARNING:
Too many gooseberries can cause stomach upset, especially in the uninitiated.

DESCRIPTION: Erect to sprawling shrubs with spiny branches. Leaves alternate, maple leaf–like, 3–5-lobed, about 1–2" (2.5–5 cm) wide. Flowers whitish to pale greenish-yellow, about $^1/_4$–$^3/_8$" (0.75–1 cm) long, tubular, with 5 small, erect petals and 5 larger, spreading sepals, single or paired in leaf axils, in May to June. Fruits smooth, purplish, ripe berries, about $^3/_8$" (1 cm) across.

Northern gooseberry (*R. oxyacanthoides*) has very spiny branches, its leaves are usually glandular-hairy beneath, and its stamens and petals are almost equal. It grows in moist woods in plains, foothills and montane zones from the Yukon and NWT to Wyoming.

White-stemmed gooseberry (*R. inerme*) has relatively spine-free older branches, glandless leaf surfaces and stamens 1.5–2 times as long as the petals. It grows in foothill and montane forests from BC and Alberta to New Mexico.

White-stemmed gooseberry

Fruit of northern gooseberry

Flowers of bristly black currant (both images)

Prickly currants

Ribes spp.

FOOD: Most currants are delicious in pies, either alone or mixed with other fruit. Bristly black currants were widely used for food, although they are sometimes described as insipid. Dried currants were occasionally included in pemmican and they make a tasty addition to bannock, muffins and breads. Currants may be boiled to make tea or mashed in water and fermented with sugar to make wine. The leaves, branches and inner bark of bristly black currant produce a menthol-flavored tea, sometimes called 'catnip tea.' To make this tea, the spines were singed off and the branches (fresh or dried) were steeped in hot water or boiled for a few minutes. Bristly fruits were rolled on hot ashes to singe off their soft spines. Both red currant jelly and the juice of green currants are said to be correctives for spoiled food (especially high meat), but this use is not recommended. For more information, see gooseberries (pp. 52–53) and black and red currants (pp. 56–57).

MEDICINE: Tea made with bristly black currant was used to treat colds, and catnip tea (above) was taken to relieve cold symptoms and diarrhea. Black currants were boiled with sugar to make syrups and lozenges for treating sore throats. A spoonful of black currant jelly in a cup of hot water has a similar soothing effect. Red currant jelly, on the other hand, has been recommended as a soothing salve for burns, said to relieve pain and reduce blistering (especially if it can be applied immediately). Black currant seeds contain gamma-linoleic acid, a fatty acid that has been used in the treatment of migraine headaches, menstrual problems, diabetes, alcoholism, arthritis and eczema. For more information, see gooseberries.

WARNING:
The spines of bristly black currant can cause serious allergic reactions in some people. Eating too many currants will cause stomach upset, especially in the uninitiated.

OTHER USES: Some native peoples considered the spiny branches (and by extension, the fruit) to be poisonous. However, these dangerous shrubs could also be useful—their thorny branches were thought to ward off evil forces.

DESCRIPTION: Erect to spreading, deciduous shrubs, 1½–5' (0.5–1.5 m) tall, with spiny, prickly branches and alternate, 3–5-lobed, maple leaf–like leaves. Flowers reddish to maroon, saucer-shaped, about ¼" (6 mm) wide, forming hanging clusters of 7–15, in May to August. Fruits ¼" (5–8 mm) wide berries, bristly with glandular hairs.

Bristly black currant (*R. lacustre*) has glandless, mostly hairless leaves and purple-black berries on slender stalks. It grows in moist foothill to alpine sites from the Yukon and NWT to Colorado and Utah.

Mountain prickly currant (*R. montigenum*) has leaves that are glandular-hairy on both sides, and its berries are bright red. It grows on rocky montane, subalpine and alpine slopes from southern BC to New Mexico.

Fruit of bristly black currant (both images)

Northern black currant

Black & red currants

Ribes spp.

FOOD: Raw currants tend to be very tart, but these common shrubs can provide a safe source of food in an emergency. Wax currants have been described as tasteless, bitter and similar to dried crabapples. Golden currant is said to be one of the most flavorful currants. Some species have a skunky smell and flavor when raw but are delicious cooked. All are high in pectin and make excellent jams and jellies that are delicious with meat, fish, bannock or toast. Some have been used to flavor liqueurs or fermented to make delicious wines. Some tribes ate *Ribes* flowers as a nibble or cooked the young leaves and ate them with raw fat. For more information, see gooseberries (pp. 52–53) and prickly currants (pp. 54–55).

MEDICINE: Some tribes ate wax currants as a strengthening tonic and as a treatment for diarrhea. In Europe, currant juice was taken to relieve arthritic pain. In the north, currants were believed to reduce the chance of heart disease caused by a high-fat diet. The roots were sometimes boiled to make medicinal teas for treating kidney problems. For more information, see gooseberries.

OTHER USES: Some native peoples believed that northern black currant had a calming effect on children, so sprigs were often hung on baby carriers. Currant shrubs growing by lakes were an indicator of fish, and in some legends, when currants dropped into the water they were transformed into fish.

DESCRIPTION: Erect to ascending, deciduous shrubs without prickles, but often dotted with yellow, crystalline resin-glands, with a sweet tomcat odor. Leaves alternate, 3–5-lobed, usually rather maple leaf–like. Flowers small (about $^1/_4$–$^3/_8$" [5–10 mm]), with 5 petals and 5 sepals, borne in elongating clusters, in spring. Fruits tart, juicy berries (currants), often speckled with yellow resin-dots or bristling with stalked glands.

Wax currant

WARNING: In the late 1700s, the Arctic explorer Samuel Hearne reported that large quantities of northern black currants could cause severe diarrhea and vomiting, but that these results would not occur if currants were mixed with cranberries. Native peoples warned that too many wax currants cause illness. A related species, sticky currant (*Ribes viscosissimum*), which has glandular-sticky shrubs and black berries, is reported to cause vomiting, even in smaller quantities.

Northern black currant (*R. hudsonianum*) has elongated clusters of 6–12 saucer-shaped, white flowers or shiny, resin-dotted, black berries. Its relatively large (2–2³/₄" [5–7 cm]), maple leaf–like leaves have resin dots on the lower surface. It grows in moist foothill and montane woods from Alaska to Wyoming.

Wax currant (*R. cereum*) has small 2–4-flowered clusters of pinkish to greenish flowers and red, glandular-hairy berries. Its small (¹/₂–1⁵/₈" [1–4 cm]), shallowly lobed, fan-shaped leaves are glandular on both sides. It grows on dry slopes in plains to montane zones from BC to New Mexico.

Golden currant (*R. aureum*) is named for its showy bright yellow flowers, not its smooth fruits, which range from black to red and sometimes yellow. Its leaves have 3 widely spreading lobes and few or no glands. It grows in plains and foothill sites from Alberta to New Mexico.

Flowers of northern black currant (top); Fruit of wax currant (above)

Wax currant

Manitoba maple keys

Acer spp.

FOOD: Maples have sweet spring sap, well known for its use in making syrup and sugar. Some tribes gathered sap from large Manitoba maple trees in early spring, before the buds had swollen. A wooden peg was set in a small wedge in the trunk bark and the sap that ran down it was collected in a birch-bark pail or a sac made from a deer stomach. This watery sap was reduced to syrup by throwing in hot stones to make it boil, then freezing it overnight and removing the ice. Some tribes made a favorite chewy candy by taking thin shavings from the inside of animal hides and mixing them in maple syrup. Maple sap can be used straight from the tree as a drink or as a cooking liquid, and it can also be fermented to make wine or vinegar. The sweet inner bark of Rocky Mountain maple was gathered in spring and used to make wine. Some sources report that de-winged maple seeds were cooked in milk and butter as a vegetable, but that they are not very palatable. Others say that sprouted seeds were used. Neither use is recommended (see Warning).

MEDICINE: Branches of Rocky Mountain maple were used to make teas that were taken as medicine and used as washes to reduce swellings and heal snake bites. Sometimes sticks of saskatoon (p. 69) were added, and the tea was given to new mothers to heal internal injuries and stimulate milk flow. Tea made by boiling bits of wood and bark was taken to cure sickness caused by being around dead bodies. The inner bark of Manitoba maple, collected after the spring sap had stopped flowing, was used to make a tea for inducing vomiting.

OTHER USES: Rocky Mountain maple wood is very strong when dried, and it was used to make arrows, snowshoes, bows, spoons, handles, hoops and cradle frames. Large burls or knots on the lower trunks of Manitoba maple trees were used to make bowls, dishes and drums, and the small branches were sometimes made into pipe stems. Maple wood produces hot, long-lasting fires with good coals for cooking meat, burning incense and making spiritual medicines. Both maples are fast growing and have colorful autumn leaves, so they are popular as ornamental shrubs and trees.

WARNING:
Some maple seeds are reported to be poisonous. Ingestion of wilted maple leaves may break down red blood cells and cause acute anemia in horses.

DESCRIPTION: Deciduous shrubs or small trees with opposite branches and hanging, V-shaped pairs of winged seeds (keys) about 1" (2.5 cm) long.

Rocky Mountain maple (*A. glabrum* var. *douglasii*) is a clumped shrub with typical maple leaves $^3/_4$–4" (2–10 cm) wide that turn orange in autumn. Its yellowish-green, $^3/_8$" (8 mm) wide flowers produce wrinkled keys. Rocky Mountain maple grows in moist foothill and montane sites from BC and Alberta to New Mexico.

Manitoba maple or **box-elder** (*A. negundo*) is a small tree with either male or female flowers. Its ash-like leaves are divided into 3–5 leaflets and turn yellow in autumn. It is a widespread garden escape, but in the Rockies it grows naturally along streams from Idaho to New Mexico.

Rocky Mountain maple (all images)

Bush-cranberries

Viburnum spp.

American bush-cranberry (top)
High bush-cranberry (above)

FOOD: These berries are best picked in autumn, after they have been softened and sweetened by freezing. Some people compare their fragrance to that of dirty socks, but the flavor is good. The addition of lemon or orange peel is said to eliminate the odor. Raw bush-cranberries can be very sour and acidic (much like true cranberries, pp. 98–99), but many native peoples ate them, chewing the fruit, swallowing the juice and spitting out the tough skins and seeds. They were also eaten with bear grease, or in an early year they could be mixed with sweeter berries such as saskatoons (p. 69). Bush-cranberries are an excellent winter-survival food, because they remain on branches all winter and are much sweeter after freezing. Some tribes ate the boiled berries mixed with oil and occasionally this mixture was whipped with fresh snow to make a frothy dessert. Today, bush-cranberries are usually boiled, strained (to remove the seeds and skins) and used in jams and jellies. These preserves usually require additional pectin, especially after the berries have been frozen. Early settlers found that berries that were not fully ripe (not yet red) jelled without pectin. Bush-cranberry juice can be used in cold drinks or fermented to make wine, and the berries can also be steeped in hot water to make tea. Their large stones and tough skins limit use in muffins, pancakes and pies.

MEDICINE: Bush-cranberries are rich in vitamin C. The bark is said to have a sedative effect and to relieve muscle spasms, so it has been widely used to treat menstrual pains, stomach cramps and sore muscles. Bush-cranberry bark has also been used to relieve asthma, hysteria and convulsions, and it was sometimes given to women threatened with miscarriage (to stop contractions). There is some controversy over bush-cranberry's ability to relieve cramps, but some sources report that pharmacological research supports this use. Commercial 'crampbark' was sometimes really the bark of mountain maple (*Acer spicatum*) because of mistakes made by collectors.

WARNING:
Some sources classify raw bush-cranberries as poisonous, while others report that they were commonly eaten raw by native peoples. A few berries may be harmless, but large quantities cause vomiting and cramps, so it is probably best to cook before eating. One source reports that the bark tea caused nausea and vomiting.

OTHER USES: A few people smoked bush-cranberry bark. The berries give a lovely reddish-pink dye. The acidic juice was also used as a mordant to set some dyes. American bush-cranberry is used as a garden ornamental.

DESCRIPTION: Deciduous shrubs with opposite, 3-lobed leaves, 1¼–4¾" (3–12 cm) long. Flowers white, small, 5-petaled, forming flat-topped clusters, in May to July. Fruits juicy, strong-smelling, red to orange, berry-like drupes ½–⅝" (1–1.5 cm) long, with a single flat stone.

American bush-cranberry or **Pembina** (*V. opulus*), which is sometimes called high bush-cranberry, is a 3½–13' (1–4 m) tall shrub with relatively deeply cut leaves. Its flower clusters have a showy outer ring of large (½–1" [12–25 mm] wide), sterile flowers. American bush-cranberry grows in foothill and montane forests from southern BC and Alberta to Wyoming.

High bush-cranberry, mooseberry or **squashberry** (*V. edule*), which is sometimes called low bush-cranberry, is a 2–7' (0.5–2 m) tall shrub whose small, relatively inconspicuous flower clusters lack showy, sterile blooms. Its distinctive, musty smell may announce its presence before it is actually seen. It grows in shady foothill, montane and subalpine sites from Alaska to Colorado.

American bush-cranberry (both photos)

Elderberries

Sambucus spp.

Red elderberry

FOOD: Generally, raw elderberries are considered inedible and cooked berries edible (see Warning), but some tribes are said to have eaten large quantities fresh from the bush. Cooking or drying destroys the rank-smelling, toxic compounds. Most elderberries were steamed or boiled or were dried for winter use. Sometimes clusters were spread on beds of pine needles in late autumn and then covered with more needles and eventually with an insulating layer of snow. These caches were easily located by their bluish-pink stain in the snow. Only small amounts were eaten at a time—just enough to get a taste. Sometimes elderberries were steamed with black hair lichens (p. 232) for flavoring. Today, they are used in jams, jellies, syrups, preserves, pies and wine. Because they contain no pectin, they are often mixed with tart, pectin-rich fruits such as crabapples. Elderberries are also used to make teas or cold drinks and to flavor some wines (e.g., Liebfraumilch) and liqueurs (e.g., Sambuca). Red elderberry juice was sometimes used to marinate salmon prior to baking. The flowers can be used to make tea or wine, and in some areas flower clusters were popular dipped in batter and fried as fritters or stripped from their relatively bitter stalks and mixed into pancake batter.

MEDICINE: Teas made from the bark and leaves were used in washes for healing eczema, sores and rashes and were drunk to treat a wide range of ailments, but ingestion is not recommended (see Warning). Hot elderberry-flower tea stimulates perspiration and has been used to alleviate cold and flu symptoms and to reduce fevers. Elderberry flowers also stimulate urination and bowel movements, so they have been used in 'slimming' pills and

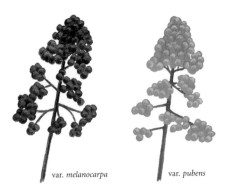

var. *melanocarpa*   var. *pubens*

WARNING:
The stems, bark, leaves and roots contain poisonous cyanide-producing glycosides (especially when fresh), which cause nausea, vomiting and diarrhea, but the ripe fruits and flowers are edible. The seeds, however, contain toxins that are most concentrated in red-fruited species. Many sources classify red-fruited elderberries as poisonous and black- or blue-fruited species as edible.

laxatives. Dried elderberry flowers have been used to make washes for treating sores, blisters, hemorrhoids, rheumatism and arthritis. The inner bark tea has been used to treat epilepsy, to induce vomiting, empty the bowels and to stimulate sweating and urination. Some species are included in herbal purges. Elderberries are rich in vitamin A, vitamin C, calcium, potassium and iron. They have also been shown to contain anti-viral compounds that could be useful in treating influenza.

OTHER USES: The pithy branches were hollowed out to fashion whistles, flutes, drinking straws, blowguns, pipe stems and spigots for tapping sap from trees, but remember, they are toxic. The berries produce crimson or violet dyes. Elderberry wine, elderberries soaked in buttermilk and elderflower water have all been used in cosmetic washes and skin creams. The leaves were crushed to make mosquito repellent or boiled to make a caterpillar repellent for spraying on garden plants.

Red elderberry

DESCRIPTION: Unpleasant-smelling, 3¹/₂–10' (1–3 m) tall deciduous shrubs with pithy, opposite branches often sprouting from the base. Leaves pinnately divided into 5–9 sharply toothed leaflets about 2–6" (5–15 cm) long. Flowers white, ¹/₈–¹/₄" (4–6 mm) wide, forming crowded, branched clusters, in April to July. Fruits juicy, berry-like drupes ¹/₈–¹/₄" (4–6 mm) across, in dense, showy clusters.

S. racemosa has pyramid-shaped flower clusters and shiny fruits. It includes 2 common varieties: **black elderberry** (var. *melanocarpa*), with purplish-black fruit, and **red elderberry** (var. *pubens*), with red fruit. Both grow in moist foothill, montane and subalpine sites from BC and Alberta to New Mexico.

Blue elderberry (*S. cerulea*) has flat-topped flower clusters and dull blue fruits with a whitish bloom. It grows on foothill and montane slopes and in valleys from BC and Alberta to New Mexico.

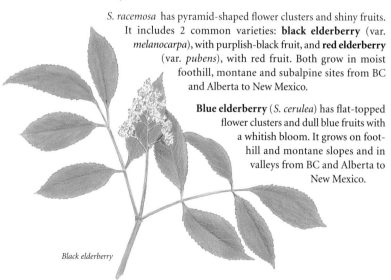
Black elderberry

Black hawthorn

Hawthorns
Crataegus spp.

FOOD: The fruits or haws of all species are edible, though they are usually rather seedy, mealy and tasteless. They were eaten fresh from the tree or dried for winter use or as an addition to pemmican. The cooked, mashed pulp (with the seeds removed) could be dried in cakes as berry-bread, which was used in soup or eaten with deer fat or marrow. Haws are rich in pectin and are usually boiled with sugar to make jams and jellies. They can also be steeped to make tea, cold drinks or juice for making wine.

MEDICINE: Hawthorn flowers and fruits are famous in herbal medicine as heart tonics, though not all species are equally effective. Studies have supported the use of hawthorn extracts as a treatment for high blood pressure associated with a weak heart, angina pectoris (recurrent pain in the chest and left arm owing to a sudden lack of blood in the heart muscle) and arteriosclerosis (loss of elasticity and thickening of the artery walls). Hawthorn is said to slow the heart rate and reduce blood pressure by dilating the large arteries supplying blood to the heart and by acting as a mild heart stimulant. However, it has a gradual, mild action and must be taken for extended periods to produce noticeable results. Hawthorn tea has also been used to treat kidney disease and nervous conditions such as insomnia. Dark-colored haws are especially high in flavonoids and have been steeped in hot water to make teas for strengthening connective tissues damaged by inflammation. The haws were sometimes eaten in moderate amounts to relieve diarrhea (some tribes considered them very constipating), and diets high in hawthorn have also been recommended in weight-loss programs. Tea made from the inner bark was used for treating dysentery.

OTHER USES: Hawthorn wood is very hard and makes excellent walking sticks. These tough, hardy shrubs have been planted on highway medians to prevent cars from swerving across into oncoming traffic. The thorns have been used as awls and fish-hooks.

WARNING:
Eye scratches from these thorns often result in blindness. Too many haws may cause nausea, vomiting and diarrhea. Hawthorns contain compounds that affect blood pressure and heart rate. People with heart conditions should use these shrubs only under the guidance of a doctor. Children and pregnant or nursing women should avoid using hawthorns.

DESCRIPTION: Deciduous shrubs or small trees with strong, straight thorns. Leaves alternate, generally oval, with a wedge-shaped base. Flowers whitish, 5-petaled, unpleasant-smelling, forming showy, flat-topped clusters, in May to June. Haws small, pulpy, red to purplish pomes (tiny apples), containing 1 nutlet.

Black hawthorn (*C. douglasii*) is most common and was widely used in the Rockies. It has $^1/_2$–$^3/_4$" (1–2 cm) long thorns, toothed to shallowly lobed leaves and black, $^3/_8$" (1 cm) long haws. Black hawthorn grows on well-drained foothill, montane and subalpine slopes from BC and Alberta to Wyoming.

River hawthorn (*C. rivularis*) is similar to black hawthorn, but it usually has longer ($^3/_4$–$1^1/_4$" [2–3 cm]) thorns and narrower (1.5–2 times longer than wide), unlobed leaves. It grows in foothill valleys from Wyoming to New Mexico.

Red hawthorn (*C. columbiana*) has much longer ($1^5/_8$–$2^3/_4$" [4–7 cm]) thorns and red haws. It grows on open plains and foothill slopes in BC, Alberta, Idaho and Montana.

Black hawthorn (both photos)

Choke cherry
Prunus virginiana

FOOD: Choke cherries were among the most important berries for many tribes. They were collected after the first frost (which makes them sweeter) and were dried or cooked, often as an addition to pemmican or stews. Large quantities were collected, pulverized with rocks, formed into patties (about 6" [15 cm] in diameter and 3/4" [2 cm] thick) and dried for winter use. Today, choke cherries are used to make jelly, syrup, sauce and wine. The raw cherries are sour and astringent, so they cause a puckering or choking sensation when they are eaten—hence the name 'choke cherry.' After they have been cooked or dried, however, they are much sweeter and lose their astringency.

MEDICINE: Dried, powdered cherry flesh was taken to improve appetite and relieve diarrhea and bloody discharge of the bowels. The inner bark was used to make medicinal teas for treating diarrhea, heart and lung problems, stimulating sweating, reducing fevers and expelling worms. It was also added to drops, syrups and medicines for soothing coughs.

OTHER USES: Crushed leaves or thin strips of bark will kill insects in an enclosed space.

DESCRIPTION: Deciduous shrub or small tree with smooth, grayish bark marked with small horizontal slits (slightly raised pores called lenticels). Leaves 1 1/4–4" (3–10 cm) long, finely sharp-toothed, with 2–3 prominent glands near the stalk tip. Flowers creamy white, 3/8–1/2" (10–12 mm) across, 5-petaled, forming bottlebrush-like clusters 2–6" (5–15 cm) long, in May to June, producing hanging clusters of 1/4" (6–8 mm) wide, dark red to black cherries. Choke cherry grows in dry to moist, open sites in plains to montane zones from southern NWT to New Mexico.

WARNING: All parts of the choke cherry (except the flesh of the fruit) contain the poison hydrocyanic acid. There are reports of children dying after eating large amounts of choke cherries without removing the stones. Cooking or drying destroys the cyanide. Choke cherry leaves and twigs are poisonous to animals.

Red cherries

Prunus spp.

FOOD: These wild cherries may be eaten raw, as a tart nibble, but the cooked or dried fruit is much sweeter and additional sugar further improves the flavor. Usually these cherries are cooked, strained and made into jelly, syrup, sauce or wine. The fruit seldom contains enough natural pectin to make a firm jelly, so pectin must be added. Although wild cherries are small, compared to domestic varieties, they are still collected in large quantities. The cherry flesh can be used in pies, muffins and pancakes, but pitting such small fruits is a tedious job.

OTHER USES: Crushed leaves or thin strips of bark will kill insects in an enclosed space. Cherry wood is very hard. Seasoned branches make excellent canes and walking sticks.

DESCRIPTION: Shrubs or small trees with raised horizontal pores (lenticels) on reddish-brown branches. Leaves 1¼–4" (3–10 cm) long, with 2 small glands near the stalk tip. Flowers white, about ³/₈" (1 cm) across, 5-petaled, forming small, flat-topped clusters, in May to June.

Pin cherry (*P. pensylvanica*) has slender-pointed, sharp-toothed leaves and bright red cherries ¹/₈–³/₈" (4–8 mm) long. It grows in plains to subalpine zones from southern NWT to Colorado, east of the Continental Divide.

Pin cherry (both photos)

WARNING:
Cherry leaves, bark, wood and seeds (stones) contain hydrocyanic acid and therefore can cause cyanide poisoning. The flesh of the cherry is the only edible part. The stone should always be discarded, but cooking or drying will destroy the toxins. Cherry leaves and twigs are poisonous to browsing animals.

Bitter cherry (*P. emarginata*) has blunt-tipped, blunt-toothed leaves and red to black cherries ³/₈–¹/₂" (8–12 mm) long. It grows in foothills zones from BC to Utah, mainly west of the Continental Divide.

Creambush oceanspray

Oceansprays

Holodiscus spp.

FOOD: The small, dry, flattened fruits of both species were eaten raw, cooked or dried by some tribes. Mountainspray roots were used to make tea.

MEDICINE: The leaves or dried seeds were boiled to make medicinal teas for treating influenza, and inner-bark tea was used as an eyewash. The bark tea was used as a tonic for convalescents and athletes, and it was taken to treat internal bleeding, diarrhea, stomach upset, flu and colds. Some tribes used the flowers to relieve diarrhea and used the leaves in poultices for soothing sore feet and chapped lips.

OTHER USES: The wood is extremely hard—other common names included 'arrow-wood' and 'ironwood'—and straight young shoots were used to make arrow, spear and harpoon shafts. Larger branches were fashioned into a variety of tools and utensils, including digging sticks, bows, tipi pins, fish clubs, drum hoops and canoe paddles. The wood was reputed not to burn, and for this reason it was used to make roasting tongs. Pioneers made wooden nails, and the Thompson tribe made breast-plates and other types of armor from oceanspray wood.

DESCRIPTION: Slender, deciduous shrubs with arching, slightly angled branches. Leaves oval, coarsely toothed to shallowly lobed. Flowers creamy white, $1/4$" (5 mm) across, 5-petaled, forming feathery, branched clusters, in June to August, that often persist for at least 1 year. Fruits hairy seed-like achenes, about $1/8$" (2 mm) long.

Creambush oceanspray (*H. discolor*) has $1^5/8$–$2^3/4$" (4–7 cm) long leaf blades, with pale, curly-hairy lower surfaces. It grows on foothill and montane slopes in southern BC, Idaho and Montana.

Mountainspray (*H. dumosus*) has smaller ($1/2$–$3/4$" [1–2 cm] long) blades, with glandular or straight-hairy lower surfaces. It grows on foothill and montane slopes from Idaho and Wyoming to New Mexico.

Saskatoon

Amelanchier alnifolia

ALSO CALLED: Juneberry • serviceberry • shadbush.

FOOD: These sweet fruits were eaten fresh, dried like raisins or mashed and dried. Some tribes steamed saskatoons in spruce bark vats filled with alternating layers of red-hot stones and fruit. The cooked fruit was mashed, formed into cakes and dried over a slow fire. These cakes could weigh as much as 15 lb (7 kg). Dried saskatoons were mixed with meat and fat to make pemmican or were added to soups and stews. Today, they are used in pies, pancakes, puddings, muffins, jams, jellies, sauces, syrups and wine, much like blueberries.

MEDICINE: Saskatoon juice was taken to relieve stomach upset and was boiled to make ear drops. Green or dried berries were used to make eye drops, and root tea was taken to prevent miscarriage. Tea made by boiling the inner bark was given to women to help them pass afterbirth. A sharpened saskatoon stick, driven deeply into the swollen ankle of a horse, was used to drain blood and other liquids.

OTHER USES: These hard, strong, straight branches were a favorite material for making arrows, spears and pipe stems. They were also used for canes, canoes (cross-pieces), basket rims, tipi stakes and tipi closure pins. The berry juice provided a purple dye. Saskatoons are excellent ornamental shrubs. They are hardy and easily propagated, with beautiful, white blossoms in spring, delicious fruit in summer and colorful, often scarlet, leaves in autumn.

DESCRIPTION: Shrub or small tree with smooth, dark gray bark, often forming thickets. Leaves oval to nearly round, 3/4–2" (2–5 cm) long, coarsely toothed on the upper half, yellowish-orange to reddish-brown in autumn. Flowers white, about 3/4" (2 cm) across, forming short, leafy clusters near the branch tips, in April to July. Fruits juicy, berry-like pomes, purple to black with a whitish bloom, 1/4–1/2" (6–12 mm) across. Saskatoon grows in open woods and hillsides in plains, foothills and montane zones from Alaska to Colorado.

WARNING:
The leaves and pits contain poisonous cyanide-like compounds. Cooking or drying destroys these toxins.

Thimbleberry
Rubus parviflorus

FOOD: Thimbleberries can be tasteless, tart or sweet, depending on the season and habitat. They are fairly coarse and seedy, and the ripe fruits do not fall free from their receptacles, which makes them difficult to pick. Thimbleberries are so common that most tribes used them for food along the trail. Thimbleberries do not dry or keep well, so they were usually eaten fresh, rather than stored for winter. Young shoots can be peeled and eaten raw or cooked. The flowers make a pretty addition to salads. The fresh or dried leaves can be used to make tea (see Warning).

MEDICINE: The leaf or root tea was said to relieve diarrhea (when served clear) and to calm upset stomach and vomiting (when mixed with milk). Some tribes ate these berries to treat chest disorders.

OTHER USES: The large leaves were widely used as plates, containers, basket liners and toilet paper.

DESCRIPTION: Deciduous shrub without prickles, often forming dense thickets. Leaves maple leaf–like, 2–8" (5–20 cm) wide, with long, glandular stalks. Flowers white, 1–2" (2.5–5 cm) across, with 5 broad, 'crinkled' petals, forming small, flat-topped clusters at branch tips, in May to July. Fruits shallowly domed, ⁵/₈–³/₄" (15–20 mm) wide, raspberry-like clusters of dull red druplets. Habitats include open or wooded foothill and montane sites from BC and Alberta to New Mexico.

WARNING:
Wilted leaves can be toxic. Only fresh or completely dried leaves should be used to make tea, but even then the tea should be used in moderation, because extended use can irritate the stomach and bowels.

Raspberries

Rubus spp.

FOOD: This colorful, delicious fruit is popular fresh from the branch or added to pies, cakes, puddings, cobblers, jams, jellies, juices and wines. Because the cupped fruit clusters drop from the receptacle when ripe, they are soft and easily crushed. Tender, young shoots, peeled of their bristly outer layer, are edible raw or cooked. Fresh or dried leaves have been used to make tea (see Warning), and the flowers can make a pretty addition to salads.

MEDICINE: Raspberry-leaf tea has traditionally been given to women before, during and after childbirth to prevent miscarriage, reduce labor pains and increase milk flow. It has also been used to slow excessive menstrual flow. Experiments have validated its use as an antispasmodic for treating painful menstruation. Raspberry leaf contains fragarine, a compound that acts both as a relaxant and a stimulant on the muscles of the uterus. Unfortunately, studies

Wild red raspberry (both photos)

on the effectiveness of raspberry leaves for treating 'female problems' have often proved inconclusive or contradictory. Raspberry leaf tea and raspberry juice boiled with sugar have been gargled to treat mouth and throat inflammations.

DESCRIPTION: Prickly shrubs with branches (canes) that live 2 years. Leaves divided into 3–5 doubly saw-toothed leaflets. Flowers white, about ³/₈" (1 cm) across, forming small clusters near the branch tips, in June to July. Fruits juicy druplets, in clusters that are about ³/₈" (1 cm) across (raspberries).

WARNING:

Wilted leaves can be toxic. Only fresh or completely dried leaves should be used to make tea, but even then the tea should be taken in moderation, because extended use can irritate the stomach and bowels.

Wild red raspberry (*R. idaeus*) has red fruit and slender prickles. It grows in foothills and montane zones from Alaska to New Mexico.

Black raspberry (*R. leucodermis*) has dark purplish to black fruit and flattened, downward-hooked thorns. It grows in foothills and montane zones from southern BC to Wyoming.

Wild roses

Rosa spp.

FOOD: Most parts of rose shrubs are edible. The hips remain on the branches throughout winter, so they are available when most other fruits are gone. Hips can be eaten fresh or dried or used in tea, jam, jelly, syrup and wine. Usually only the fleshy outer layer is eaten (see Warning). Because they are so seedy, some tribes considered rose hips famine food. Rose petals may be eaten alone as a trail nibble, added to salads, teas, jellies and wines or candied. Rose leaves, roots and peeled twigs have also been used in teas. Buds, young shoots and young leaves may be eaten raw or cooked.

MEDICINE: Rose hips are rich in vitamins A, B, E and K, and 3 hips can contain as much vitamin C as an orange. During World War II, when oranges could not be imported, British and Scandinavian people collected hundreds of tons of rose hips to make syrup. The vitamin C content of fresh hips varies greatly, but that of commercial 'natural' rose hip products can fluctuate even more. Stem or root bark tea was taken to relieve diarrhea or stomach upset and to reduce labor pain. It was also used as an eyewash for treating snow-blindness. Root decoctions were used in hot compresses for reducing swelling, were gargled to treat mouth bleeding, tonsillitis and sore throat or were mixed with sugar to make a syrup for soothing sore throats. The leaves were boiled to make a wash for strengthening babies. Rose petals were taken to relieve colic, heartburn, headaches and mouth sores. They were also ground and mixed with grease to make a salve for mouth sores or mixed with wine to make a medicine for relieving earaches, toothaches and uterine cramps. Cooked seeds were eaten to relieve sore muscles.

Prickly rose's fruit and flowers

OTHER USES: Dried rose petals have a lovely fragrance and have been used in potpourri. The inner bark was sometimes smoked like tobacco (*Nicotiana tabacum*), and the roots were boiled to make hair rinse. Pithy stems were occasionally used for pipe stems or arrow shafts. Rose sprigs were hung on cradle boards to keep ghosts away from babies, and on the walls of haunted houses and in graves to prevent the dead from howling. Alberta, Iowa and North Dakota all have wild roses as their floral emblems.

DESCRIPTION: Thorny to prickly, deciduous shrubs with leaves pinnately divided into about 5–7 oblong, toothed leaflets. Flowers light pink to deep rose, 5-petaled, fragrant, forming branched clusters, in June to August. Fruits scarlet to purplish, round to pear-shaped, berry-like hips, ⅝–1¼" (1.5–3 cm) long, with a fleshy outer layer enclosing many stiff-hairy achenes.

Prickly rose (*R. acicularis*) has bristly prickly branches and small clusters of 2–2¾" (5–7 cm) wide flowers or ½–¾" (1–2 cm) long hips. It grows in a wide range of sites in the plains to subalpine zones from Alaska to New Mexico.

Prairie rose (*R. woodsii*) has well-developed thorns at its joints, small clusters of 1⅝–2½" (4–6 cm) wide flowers and ¼–⅜" (6–10 mm) long hips. It grows in many habitats from the

Prairie rose

plains to subalpine zones from the southern Yukon and NWT to Colorado and Utah.

Nootka rose (*R. nutkana*) has well-developed thorns at its joints, large (about 3" [8 cm] wide), mostly single flowers and ⅝–¾" (1.5–2 cm) long hips. It grows in a wide range of habitats in foothills and montane zones from southern BC to Colorado and Utah.

WARNING:
The dry inner 'seeds' (achenes) are not palatable and their sliver-like hairs can irritate the digestive tract and cause 'itchy bum.' All members of the Rose family have cyanide-like compounds in their seeds, destroyed by drying or cooking.

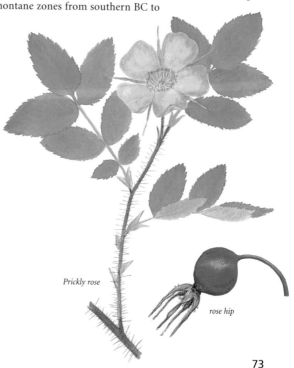

Prickly rose

rose hip

Mountain-ashes
Sorbus spp.

Western mountain-ash

FOOD: These bitter fruits can be eaten raw, cooked or dried, but many tribes considered them inedible. The green berries are too bitter to eat, but the ripe fruit, mellowed by repeated freezing, is said to be quite tasty. Mountain-ash fruit is usually cooked and sweetened. It has been used to make jams, jellies, pies, ale and a bittersweet wine. The berries have also been dried and ground into flour. In northern Europe, this flour was fermented and used to make a strong liquor.

MEDICINE: Native peoples boiled the peeled branches or inner bark to make teas for treating back pain, colds, headaches, sore chests and internal bleeding (perhaps associated with tuberculosis). The branches were boiled and then the steam was inhaled to relieve headaches and sore chests. The berries can be used to make a juice rich in vitamin C. The astringent berry tea has been used as a gargle for relieving sore throats and tonsillitis. European mountain-ash fruit was used to make medicinal teas for treating indigestion, hemorrhoids, diarrhea and problems with the urinary tract, gall bladder and heart. Teas made from its bark and leaves were used to treat malarial fevers, ulcers, hemorrhoids and sores.

OTHER USES: European mountain-ash is a popular ornamental tree, and the native mountain-ashes make attractive garden shrubs, easily propagated from seed sown in autumn. The scarlet fruit clusters attract many birds. Some native peoples rubbed the berries into their scalps to kill lice and treat dandruff.

DESCRIPTION: Clumped, deciduous shrubs or trees with pinnately divided, sharply toothed leaves. Flowers white, about $3/8$" (1 cm) across, 5-petaled, forming flat-topped, $3^1/2$–6" (9–15 cm) wide clusters in June to July. Fruits berry-like pomes, about $3/8$" (1 cm) long. Native species are usually $3^1/2$–13' (1–4 m) tall.

WARNING:
The fruits are said to be somewhat toxic, because they contain cyanide-related compounds and parascorbic acid (reported to be a cancer-causing compound). Cooking neutralizes these toxins.

Western mountain-ash (*S. scopulina*) has sticky twigs, orange to scarlet fruit and 9–13 shiny green, pointed leaflets, with teeth almost to the base. It grows in foothills to subalpine zones from northern Canada to New Mexico.

Sitka mountain-ash (*S. sitchensis*) has non-sticky twigs, crimson to purplish fruit and 7–11 dull bluish-green, round-tipped leaflets without teeth near the base. It grows in foothills to subalpine zones from BC and Alberta to Montana.

European mountain-ash or **Rowan tree** (*S. aucuparia*) is a widely planted ornamental tree, 16–40' (5–12 m) tall, with more than 13 leaflets per leaf.

Sitka mountain-ash

Western mountain-ash

75

Oregon-grapes

Mahonia spp.

FOOD: These sour, juicy berries have been eaten raw or used to make jelly, jam or wine. Mashed with sugar and served with milk, they are said to make a tasty dessert. A refreshing drink was made with mashed berries, sugar and water. The sweetened juice is said to taste much like grape juice. Berry production can vary greatly from year to year, and the fruits are sometimes rendered inedible by grub infestations. Very young leaves have been used as a trail nibble or salad vegetable, and the young leaves of tall Oregon-grape were sometimes simmered until tender.

MEDICINE: The roots contain the alkaloid berberis, which stimulates involuntary muscles. Root tea was used to aid in delivery of the afterbirth and to relieve constipation. It was also taken as cough medicine, though the stimulative effects of berberis would seem more likely to increase, rather than reduce, coughing. The crushed plants and roots have anti-oxidant, antiseptic and anti-bacterial properties. They were used to make medicinal teas, poultices and powders for treating gonorrhea and syphilis and for healing wounds and scorpion stings. The leaf tea was used as a tonic and/or contraceptive and as a treatment for kidney troubles, stomach troubles, diarrhea, dysentery, rheumatism, skin problems, such as psoriasis and eczema, and persistent problems of the uterus.

Tall Oregon-grape

OTHER USES: Boiled, shredded root bark produces a brilliant yellow dye. Tall Oregon-grape is often planted as an ornamental. Oregon-grape is the state flower of Oregon.

Creeping Oregon-grape

WARNING:
High doses of Oregon-grape can cause nose bleeds, skin and eye irritation, shortness of breath, sluggishness, diarrhea, vomiting, kidney inflammation and even death. Pregnant women should not use this plant, because it may stimulate the uterus.

DESCRIPTION: Low, wintergreen shrubs with leathery, holly-like leaves pinnately divided into spiny-edged leaflets, red or purple in winter. Flowers yellow, about ⅝" (1 cm) across, forming elongated clusters, in April to June. Fruits juicy, grape-like berries, about ⅝" (1 cm) long, purplish-blue with a whitish bloom.

Creeping Oregon-grape (*M. repens*) is 4–12" (10–30 cm) tall, with dull leaflets that are bluish-green (at least on the lower surface). It grows on forested foothill and montane slopes from southern BC and Alberta to New Mexico.

Tall Oregon-grape

Tall Oregon-grape (*M. aquifolium*) is a 1–5' (30–150 cm) tall shrub, whose leaves are shiny above and not whitened beneath. It grows on foothill slopes in southern BC, Idaho and Montana.

Dull Oregon-grape (*M. nervosa*) has longer leaves with 9–19 (rather than 5–9) leaflets and large (¾–1⅝" [2–4 cm] long), leathery bud scales (rather than thin, quickly shed scales less than ⅝" [1 cm] long). This 4–12" (10–30 cm) tall shrub grows on forested slopes in the foothills zone of BC and Idaho.

Creeping Oregon-grape

Smooth sumac

Rhus glabra

FOOD: The fuzzy, red fruits have been crushed, soaked in cold water and strained (removing hairs and other debris) to produce a refreshing, pink or rose-colored drink with a lemon-like flavor. It is best sweetened with sugar and served cold. Soaking in hot water produces a bitter-tasting liquid. The fruits have also been used to make 'lemon pies' and chewed as a trail nibble to relieve thirst and leave a pleasant taste in the mouth.

MEDICINE: The fruit was boiled to make a wash to stop bleeding after childbirth. The root tea was taken to treat fluid retention and painful urination. Roots were also chewed and the juice swallowed to soothe sore throats. Root bark was used in poultices for healing ulcers and open wounds. Sumac branches were used to make medicinal teas for treating tuberculosis, and the bark, steeped in hot water, produced an astringent tea that was applied to skin problems, gargled to relieve sore mouths, and taken internally to relieve diarrhea, dysentery, gonorrhea and syphilis. The bruised, moistened leaves were applied to rashes and to skin reactions associated with plant irritants such as poison-ivy. The berries, steeped in hot water, made a medicinal tea for treating diabetes, bowel problems and fevers. This tea was also used as a wash for ringworm, ulcers and skin diseases such as eczema.

OTHER USES: The leaves, bark and roots yield a yellow-tan, gray or black dye, depending on the part used and the mordant. The red, autumn leaves were sometimes rolled and smoked.

DESCRIPTION: Deciduous shrub, $3^1/_2$–10' (1–3 m) tall, often forming thickets. Leaves pinnately divided into 11–21 lance-shaped, 2–4$^3/_4$" (5–12 cm) long, toothed leaflets, bright red in autumn. Flowers cream-colored to greenish-yellow, about 3 mm across, with 5 fuzzy petals, forming dense, pyramid-shaped, 4–8" (10–20 cm) long clusters, in April to July. Fruits densely reddish-hairy, berry-like drupes, $^1/_8$–$^1/_4$" (4–5 mm) long, in persistent, fuzzy clusters. Smooth sumac grows on dry slopes in plains and foothills zones from southern BC to New Mexico.

WARNING: People who are hypersensitive to the poisonous members of this genus (e.g., poison-ivy, p. 237) may also be allergic to this 'safe' sumac.

Skunkbush

Rhus trilobata

ALSO CALLED: three-leaved sumac.

FOOD: Skunkbush fruits were eaten raw (sometimes ground with a little water) or boiled, or they were dried for later use and ground into meal. The berries and meal were often mixed with other foods, especially sugar and roasted corn. Skunkbush berries can be crushed, soaked in cold water, and strained to make a pink lemonade-like drink. Soaking in hot water produces a bitter-tasting liquid. The inner bark was also eaten.

MEDICINE: The leaves were chewed to cure stomachaches or were boiled to make a contraceptive tea that was said to induce impotence. Skunkbush leaves were also used in poultices to relieve itching. Tea made by boiling the root bark was taken to aid delivery of the afterbirth. The fruits were boiled and their oil, skimmed from the surface, was used to prevent hair loss.

OTHER USES: Skunkbush branches were split lengthwise (usually 3 times) and woven into baskets and water bottles. Sun shades or hats were woven from the smaller branches, but they were for adults only. It was said that if children placed sumac on their heads, they would stop growing. Large stems (about 6½' [2 m] long) were used to make bows or spear shafts tipped with points made of harder wood or iron. Twigs from the east side of the shrub were woven into the form of an owl, which was hung on the west side of the tipi smoke hole. The owl, moving up and down in the smoke, was used to scare children and make them behave. Making an owl was a sure sign that it would rain. The leaves were boiled to produce a black dye for baskets, leather and wool, and the ashes were used to set dyes.

WARNING: People who are hypersensitive to the poisonous members of this genus (e.g., poison-ivy, p. 237) may also be allergic to this 'safe' sumac.

DESCRIPTION: Strong-smelling, 3½–6½' (1–2 m) tall shrub with bright green leaves divided into 3 broad-tipped, lobed leaflets that taper to wedge-shaped bases. Flowers yellowish-green, about ⅛" (3 mm) across, with 5 fuzzy petals, forming close clusters of spikes near the branch tips, in May to July, before the leaves appear. Fruits sticky, fuzzy, reddish-orange, berry-like drupes, ¼–⅜" (6–8 mm) long, in small, fuzzy clusters. Skunkbush grows along streams in plains and foothills zones from Idaho and Alberta to New Mexico.

79

Buckbrushes

Ceanothus spp.

Snowbrush

FOOD: These shrubs are famous tea substitutes. Dried leaves and flowers were steeped for about 5 minutes to produce a pale, yellowish tea, milder and sweeter than commercial teas. Some people recommended species with resinous leaves, but others preferred less resinous species. Some tribes ate strips of the inner bark of Fendler's buckbrush for food during summer.

MEDICINE: Teas made with buckbrush and monument plant (*Frasera speciosa*) were applied externally and taken internally to relieve alarm or nervousness. The roots were used to make medicinal teas for treating inflamed tonsils, enlarged lymph nodes, non-fibrous cysts and enlarged spleens. This tea was also said to be an excellent home remedy for relieving excessive menstrual bleeding, nosebleeds, hemorrhoids, old ulcers and bleeding from vomiting and coughing. It has been recommended for heavy drinkers with inflammation of the stomach, whiskey nose and other symptoms of weak capillaries.

OTHER USES: Buckbrush tea was used as a shampoo and hair tonic. When the flowers are rubbed in water, they produce a soapy foam. The resinous leaves of some species have been used as a substitute for tobacco (*Nicotiana tabacum*). Buckbrush makes an attractive garden plant. The shrubs are difficult to transplant, so they are best propagated from seed. Fendler's buckbrush was used to induce vomiting during religious ceremonies.

DESCRIPTION: Sprawling, spicy-scented shrubs, usually in dense patches. Leaves broadly oval, finely toothed or toothless. Flowers white, tiny, fragrant, forming dense clusters at the tips of side branches, in June to August. Fruits glandular-sticky, 3-lobed capsules, $1/8$" (4–5 mm) long, ejecting 3 shiny seeds when mature.

WARNING:
People with blood disorders or allergies to aspirin should not take buckbrush, and it should be used in moderation during pregnancy.

Snowbrush or **velvety buckbrush** (*C. velutinus*) has pyramid-shaped flower clusters, and its sticky, shiny leaves are evergreen, with finely toothed, 1$^5/_8$–3" (4–8 cm) long blades and with 2 tiny (1 mm long) lobes at the stalk base. It grows on fairly open sites in foothills and montane zones from BC and Alberta to Colorado.

Red-stemmed buckbrush (*C. sanguineus*) also has pyramid-shaped flower clusters and toothed leaves, but its leaves are thinner, with 2 slender, $^1/_8$–$^3/_8$" (3–8 mm) lobes at the base, and they are soon shed. It grows on well-drained slopes in the foothills zones of southern BC, Idaho and Montana.

Fendler's buckbrush (*C. fendleri*) is a spiny shrub with small ($^1/_2$–1$^1/_4$" [1–3 cm] long), toothless leaves and small, umbrella-shaped clusters (umbels) of white flowers. It grows on dry slopes in foothills and montane zones from Wyoming to New Mexico.

Red-stemmed buckbrush

Snowbrush

Buckthorn & Cascara

Rhamnus spp.

Alder-leaved buckthorn (top); Cascara (above)

FOOD: Some sources report that these purple berries were eaten by native peoples, but this story seems unlikely given the berries' strong purgative effects.

MEDICINE: These shrubs have been used as laxatives for hundreds of years. The bark of alder-leaved buckthorn contains a laxative, but cascara, 'the all-American laxative,' has been much more widely used. Native peoples collected the bark in spring and summer, either by scraping downwards (if the patient needed a laxative) or upwards (if the patient needed to vomit). It was then dried and stored for later use. Ingesting fresh bark and berries could have very severe effects, but curing the bark for at least 1 year or using a heat treatment reduced the harshness. The dried bark was traditionally used to make medicinal teas, but today it is usually administered as a liquid extract or elixir or in tablet form. Each year 1–3 million lb (0.5–1.4 million kg) of bark are collected, mainly from wild shrubs in BC, Washington, Oregon and California. Some tribes used these plants to induce vomiting when poisons had been eaten.

DESCRIPTION: Erect or spreading shrubs with prominently veined, 2–4" (5–10 cm) long leaves. Flowers greenish-yellow, all male or all female, forming flat-topped clusters in leaf axils, in June to July. Fruits reddish to bluish-black, berry-like drupes, 1/4–3/8" (6–9 mm) long.

Alder-leaved buckthorn (*R. alnifolia*) is 3–5' (0.5–1.5 m) tall, with 2–5-flowered, stalkless flower clusters and 10–14 veined leaves. It grows in moist, shady foothill and montane sites from BC and Alberta to Wyoming.

Cascara (*R. purshiana* or *Frangula purshiana*) is taller (up to 33' [10 m]), with 8–40-flowered, stalked flower clusters and 20–24-veined leaves. It grows on wooded foothills zones in BC, Idaho and Montana.

WARNING:
These shrubs can cause severe vomiting and diarrhea. A few rare cases of mild poisoning have been reported. They are best avoided when gathering sticks for roasting hot dogs.

Red-osier dogwood

Cornus sericea

ALSO CALLED: *Cornus stolonifera.*

FOOD: These fruits are very bitter for modern-day tastes, but many native peoples collected them for food in the past. Berries with a bluish tinge are said to be the most sour. Usually the berries were gathered in late summer and autumn and eaten fresh, either alone or mashed with sweeter fruits such as saskatoons (p. 69). Occasionally, they were dried for later use. More recently, these berries were used to make a dish called 'sweet-and-sour' by mixing them with sugar and other sweeter berries. Some people separated the stones from the mashed flesh and saved them for later use. They were then eaten as a snack, like peanuts.

MEDICINE: Some tribes used the bark tea to treat digestive disorders, because of its laxative effects. The bark was also smoked to treat sickness of the lungs. The inner bark contains an analgesic, coronic acid, so it has been used as a salicylate-free pain-killer.

OTHER USES: The soft, white inner bark was dried and used for smoking, It was said to be both aromatic and pungent, and to have a narcotic effect that could cause stupefaction. Usually, it was mixed with tobacco (*Nicotiana tabacum*) or common bearberry (pp. 90–91). The inner bark from the roots was included in herb mixtures that were smoked in ceremonies. The flexible branches are often used in wreaths or woven into contrasting red rims and designs in baskets. This attractive shrub, with its lush green leaves and white flowers in spring, red leaves in autumn and bright red branches and white berries in winter, is a popular ornamental. It grows best in moist sites and is easily propagated from cuttings or by layering.

DESCRIPTION: Erect to sprawling, deciduous shrub with opposite, purple to red (sometimes greenish) branches. Leaves $3/4$–4" (2–10 cm) long, pointed, with 5–7 prominent veins per side. Flowers white, about $1/4$" (5 mm) across, forming dense, flat-topped clusters at the branch tips, in May to July. Fruits fleshy, white (sometimes bluish), berry-like drupes, $1/4$" (5–7 mm) across, containing large, flattened stones. Red-osier dogwood grows on moist sites in plains, foothills and montane zones from the southern Yukon and NWT to New Mexico.

WARNING: All parts of this shrub can be toxic, especially if they are consumed in large quantities.

Lewis' mock-orange

Philadelphus lewisii

MEDICINE: Fresh mock-orange leaves were bruised and used as poultices to heal infected breasts. Salves were made using either dried leaves or charcoal from burning the wood. Both were ground to a fine powder, which was then mixed with pitch or bear grease. These salves were used for treating sores and swellings. Mock-orange branches, with or without blossoms, were boiled to make medicinal teas for treating sore chests. These teas were also used to wash or soak skin affected by eczema and to stop bleeding of hemorrhoids.

OTHER USES: Mock-orange wood is stiff and hard. It was used to make small items such as combs, knitting needles, basket rims (for birch-bark baskets) and cradle hoods. The leaves were sometimes mashed in water and used as soap. This attractive shrub, with its dark green foliage and showy, fragrant flowers, makes a lovely addition to sunny or partly shaded gardens. Several cultivars are available from greenhouses and nurseries.

DESCRIPTION: Erect deciduous shrub, $3^1/_2$–10' (1–3 m) tall, with loosely spreading branches. Leaves opposite, 1–$2^3/_4$" (2.5–7 cm) long, with 3 main veins from the base, slightly roughened and fringed with short hairs. Flowers fragrant, white, about $1^1/_4$" (3 cm) across, with 4 broad spreading petals, forming showy clusters of 3–11 at the branch tips, in May to July. Fruits woody capsules $^1/_4$–$^3/_8$" (6–10 mm) long. Lewis' mock-orange grows on well-drained sites, often along streams in foothills and montane zones from BC and Alberta to Idaho and Montana.

Canada buffaloberry

Shepherdia canadensis

ALSO CALLED: soopolallie • soapberry.

FOOD: These small berries were collected by beating the branches over a canvas or hide and then rolling the berries into a container to separate leaves and other debris. Buffaloberries were eaten fresh, or they were boiled, formed into cakes and dried over a small fire for future use. More recently, they have been preserved by canning or freezing. Because their juice is rich in saponin, buffaloberries become foamy when beaten. They were mixed about 4:1 with water and whipped like egg-whites to make a white to salmon-colored, foamy dessert called 'Indian ice cream.' Traditionally, this dessert was beaten by hand or with a special stick with grass or strands of bark tied to one end, but these tools were eventually replaced by egg-beaters and mixers. Like egg-whites, buffaloberries will not foam in plastic or greasy containers. The foam is rather bitter, so it was usually sweetened with sugar or with other berries. Buffaloberries were also added to stews or cooked to make syrup, jelly or a sauce for buffalo steaks. Canned soapberry juice, mixed with sugar and water, makes a refreshing 'lemonade.' Although they are bitter, buffaloberries are often abundant and can be used, in moderation, as an emergency food (see Warning).

MEDICINE: The berries are rich in vitamin C and iron. They have been taken to treat flu and indigestion and have been made into a medicinal tea for relieving constipation. Canned soapberry juice, mixed with sugar and water, was said to cure acne, boils, digestive problems and gallstones. The bark tea was a favorite solution for eye troubles, and the twigs were boiled to make a laxative tea.

OTHER USES: The soapy berries were crushed or boiled and used as soap.

WARNING:
Buffaloberries contain a bitter, soapy substance, saponin, which can irritate the stomach and cause diarrhea, vomiting and cramps if consumed in large amounts.

DESCRIPTION: Deciduous shrub with opposite, ⁵/₈–2¹/₂" (1.5–6 cm) long, dark green leaves that are silvery-fuzzy beneath with star-shaped hairs and rust-colored scales. Flowers yellowish to greenish, ¹/₈" (4 mm) wide, either male or female on the same shrub, forming small clusters, in April to June, before the leaves appear. Fruits juicy, translucent, bright red to yellowish berries, ¹/₈–¹/₄" (4–6 mm) long. Canada buffaloberry grows in open woods and on streambanks in foothills, montane and subalpine zones from Alaska to New Mexico.

Silverberry
Elaeagnus commutata

ALSO CALLED: wolfwillow.

FOOD: These berries are very dry and astringent, but some northern tribes gathered them for food. Most groups considered the mealy berries famine food. Silverberries were eaten raw or cooked in soup. They were also mixed with blood and cooked, mixed with lard and eaten raw, fried in moose fat or frozen. Apparently, they make good jams and jellies. The berries are much sweeter after exposure to freezing temperatures.

OTHER USES: The bark was used to make cord, and several tribes used the nutlets inside the berries as decorative beads. The fruits were boiled to remove the flesh, and while the seeds were still soft, a hole was made through each. They were then threaded, dried, oiled and polished. If green silverberry wood is burned in a fire, it gives off a strong smell of human excrement. Some practical jokers enjoyed sneaking branches into the fire and watching the reactions of fellow campers.

DESCRIPTION: Thicket-forming shrubs with $3/4$–$2^1/2$" (2–6 cm) long, silvery leaves covered in dense, tiny, star-shaped hairs (sometimes also with brown scales beneath). Flowers strongly sweet-scented, yellow inside and silvery outside, $1/2$–$5/8$" (12–16 mm) long, borne in 2s or 3s from leaf axils, in June to July. Fruits silvery, mealy, about $3/8$" (1 cm) long, drupe-like, with a single large nutlet. Silverberry grows on well-drained, open sites in plains, foothills and montane zones from Alaska to Idaho and Montana.

WARNING:
These flowers can be detected from meters away by their sweet, heavy perfume. Some people enjoy this fragrance, but others find it overwhelming and nauseating.

Big sagebrush

Artemisia tridentata

FOOD: Although sagebrush seeds can be quite bitter, they were used for food by some people, and the seeds of big sagebrush were considered the best. The seeds were eaten raw or dried, but usually they were ground into meal and cooked in pinole, soups and stews. The volatile oils of these plants sometimes added flavor and fragrance to liqueurs.

MEDICINE: Big sagebrush tea was taken to treat colds, fevers, pneumonia and sore eyes and to ease childbirth. Hunters and athletes used it to cleanse the body before long hikes or competitions. The leaves, chewed and swallowed or boiled to make medicinal teas, were taken to relieve stomachaches, coughing, bleeding and postpartum pain and to expel intestinal parasites. Settlers also used big sagebrush to treat headaches, diarrhea, vomiting and bullet wounds. Wet leaves were used as poultices to reduce swelling and infection. Sagebrush extracts are believed to kill many types of bacteria and can be applied (with caution; see Warning) to cuts, scrapes and other skin problems to combat infection.

OTHER USES: These shrubs were often burned as firewood, especially if no other wood was available. The aromatic smoke from green, leafy branches provided a ceremonial smudge to cleanse participants of evil spirits and impurities. Leafy branches were used as switches in sweat-baths or were tied together with wire to make brooms. The leaf tea was applied to cuts on sheep (from barbed wire, etc.) to speed healing. The aromatic, volatile oils of these plants were used in hair tonics and shampoos and as moth and flea repellents. Sagebrush is the unofficial floral emblem of Nevada.

DESCRIPTION: Grayish-hairy, wintergreen shrub, about $1^1/_2$–$6^1/_2$' (0.5–2 m) tall. Leaves $^1/_2$–$^3/_4$" (1–2 cm) long, narrowly wedge-shaped, 3-toothed at the tip and tapered to the base. Flowerheads $^1/_{16}$–$^1/_8$" (2–3 mm) wide, yellow or brownish clusters of 5–8 disc florets, forming long, narrow ($^5/_8$" [1.5–7 cm] wide) clusters, in July to October. Fruits sparsely hairy seeds (achenes). Big sagebrush covers many acres of dry land in plains, foothills and montane zones from southern BC and Alberta to New Mexico.

WARNING:
Many people are allergic to big sagebrush and in rare cases it causes inflammation when applied to the skin. Some classify it as toxic, reporting that it can damage the liver and intestinal tract. It is probably best to use sagebrush for external applications only.

Rabbitbushes
Chrysothamnus spp.

FOOD: The milky sap (latex) of these shrubs contains rubbery compounds. The bark of the lower stem and roots of several species of rabbitbush was widely used as chewing gum.

MEDICINE: The roots of common rabbitbush were boiled to make a strong decoction for treating coughs, fevers, colds and old internal injuries and for easing menstrual cramps. To relieve headaches, the leaves were used to make a medicinal tea that was applied as a lotion. The leaf tea was also taken internally to reduce fevers and relieve constipation, colds and stomach problems. Mashed rabbitbush leaves were packed onto decayed teeth to relieve toothaches.

OTHER USES: The branches of rabbitbush were burned slowly to smoke hides. The leafy boughs were used to cover sweathouses and to carpet the floor. Mature flowers were boiled for at least 6 hours to produce a lemon-yellow dye. Alum was then added as a mordant, along with the wool or leather to be dyed, and boiling continued for about an hour. When dying baskets, the flowers and buds were boiled overnight

Common rabbitbush (both photos)

WARNING:
Livestock in California are reported to have been poisoned by eating common rabbitbush. The toxicity of rabbitbushes is not clear, so these shrubs should not be used in foods or internal medicines.

and the basket material was then soaked in the dye for about 12 hours. No mordant was used. Immature buds or twigs gave the dye a greenish tinge.

DESCRIPTION: Erect, densely branched shrubs, about 1–3$^{1}/_{2}$' (30–100 cm) tall, often flat-topped. Leaves linear, $^{1}/_{2}$–2$^{1}/_{2}$" (1–6 cm) long, undivided. Flowerheads about $^{1}/_{4}$" (5 mm) across, yellow, with about 5 disc florets and 5 overlapping rows of involucral bracts, forming dense clusters at branch tips, in August to October. Fruits slender seed-like achenes tipped with a tuft of white hairs (pappus).

Common rabbitbush (*C. nauseosus*) has felt-covered twigs and abruptly pointed involucral bracts. It grows on dry plains, foothill and montane slopes from southern Canada to New Mexico.

Parry's rabbitbush (*C. parryi*) has felt-covered twigs and slender, green-tipped involucral bracts. It grows on dry plains to subalpine sites in Wyoming and Colorado.

Green rabbitbush (*C. viscidiflorus*) has brittle, essentially hairless twigs. Green rabbitbush grows on dry plains, foothill and montane slopes from southern Canada to New Mexico.

Common rabbitbush (all images)

Bearberries & Manzanitas

Arctostaphylos spp.

FOOD: Bearberries are rather mealy and tasteless, but they are often abundant and remain on branches all year, so they can provide an important survival food. Many tribes used them for food. To reduce the dryness, bearberries were usually cooked with salmon oil, bear fat or fish eggs, or they were added to soups or stews. Sometimes boiled berries were preserved in oil and served whipped with snow during winter. Boiled bearberries, sweetened with syrup or sugar and served with cream, make a tasty dessert. They can also be used in jams, jellies, cobblers and pies, dried, ground and cooked in mush, or fried in grease over a slow fire until they pop like popcorn. Scalded mashed berries, soaked in water for an hour or so, produced a spicy, cider-like drink, which was sweetened and fermented to make wine. The dried leaves were used to make tea.

MEDICINE: The leaves, fresh or dried, were used in medicinal teas as tonics and for stimulating urination and cleansing the urinary tract. Studies suggest that bearberry has an antiseptic effect on the urinary tract, and it has been shown to inhibit the growth of gram-positive bacteria. It also has a mild vaso-constricting effect on the uterus and may help to relieve menstrual cramps. The tea was used to treat kidney and bladder problems, lower back pain, bronchitis, diarrhea, gonorrhea and bleeding. It can be used in sitz baths and washes to reduce inflammation and infection.

OTHER USES: The leaves and bark were widely smoked, often mixed with red-osier dogwood bark (p. 83), alder bark (p. 50), tobacco (*Nicotiana tabacum*) and other ingredients. Usually, they were toasted near a fire until crisp and brown and then crushed and smoked. Some people say that smoking bearberry leaves can cause dizziness and even unconsciousness in the uninitiated. The tannin-rich leaves have been used to tan hides. Hikers sometimes chew the berries and leaves to stimulate saliva flow and relieve thirst.

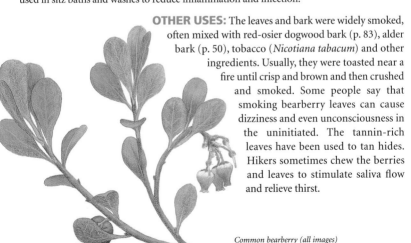

Common bearberry (all images)

DESCRIPTION: Evergreen or deciduous shrubs with clusters of nodding, white or pinkish, urn-shaped flowers and juicy to mealy, berry-like drupes containing 5 small nutlets. The genus *Arctostaphylos* contains 2 main groups of species—the bearberries, which are low, trailing to tufted shrubs found in arctic and alpine regions, and the manzanitas, which are erect or spreading, taller shrubs, of the western US.

Common bearberry (*A. urva-ursi*) is a trailing evergreen shrub with leathery, evergreen, spoon-shaped leaves ³/₈–1¹/₈" (1–3 cm) long. Its small (¹/₈–¹/₄" [4–6 mm]) flowers appear in May to June and produce bright red, ¹/₄–³/₈" (6–10 mm), mealy fruits by late summer. Common bearberry grows on well-drained, foothill, montane, subalpine and alpine sites from Alaska to New Mexico.

Alpine bearberry (*A. rubra*) has thin, veiny, deciduous, spoon-shaped leaves that often turn red in autumn. Although they are fairly insipid, these juicy red berries are probably among the most palatable fruits in the genus, but because alpine bearberry grows at high elevations and northern latitudes, it has been the least used. These low, tufted shrubs grow on montane, subalpine and alpine slopes from Alaska to New Mexico.

Alpine bearberry (top); Common bearberry (above)

WARNING:
Too many bearberries can cause constipation. Bearberry is rich in tannin and arbutin. Extended use (e.g., for more than 3 days) causes stomach and liver problems (especially in children). Bearberry should not be used by pregnant women. It can stimulate uterine contractions and reduce permeability of the placenta-uterine membrane.

Greenleaf manzanita (*A. patula*) is a much taller shrub (usually over 3' [1 m] tall) that is common on wooded plains, foothill and montane slopes in western Colorado and Utah. It has large (³/₄–2" [2–5 cm] long), egg-shaped, evergreen leaves and white to yellowish fruits. Manzanita berries are very similar to the ones of common bearberry, and they were an important food for many tribes.

Labrador teas
Ledum spp.

FOOD: These aromatic leaves, and occasionally their flowers, have been widely used for making tea. Fresh or dried leaves were steeped in boiling water, but some people prepared a stronger, darker brew by boiling them for hours or even days (see Warning). The fragrant, steeped brew was served hot, cooled or chilled, either alone or sweetened with sugar. Both species have been used for teas, but glandular Labrador tea is relatively toxic (see Warning).

MEDICINE: Labrador tea has been used to treat colds, sore throats and allergies. Although it is slightly laxative, the tea was said to soothe diarrhea and stomach upset. Some people recommended glandular Labrador tea as a relaxant. Scandinavians let a handful of crushed leaves soak in an alcoholic drink and then used the liquor as a sedative nightcap. Alcohol extracts of the leaves have been used to treat scabies, lice, chiggers and fungal skin diseases. Strong decoctions were applied to inflamed, itchy or oozing skin conditions.

OTHER USES: Dried leaves were sometimes stored in grain to repel mice and rats. They were also added to smoking mixtures, in moderation, for their mild euphoric effect and their flavor. Alcohol extracts of the leaves provided insect repellent and a remedy for insect bites.

DESCRIPTION: Evergreen shrubs, 16–32" (40–80 cm) tall. Leaves alternate, leathery, elliptic to oblong. Flowers white, about $^3/_8$" (1 cm) across, 5-petaled, forming flat-topped clusters, in June to August. Fruits round, nodding capsules, $^1/_8$–$^1/_4$" (3–5 mm) long on $^3/_8$–$^3/_4$" (1–2 cm) long stalks.

Labrador tea (top)
Glandular Labrador tea (above)

Labrador tea (*L. groenlandicum*) has leaves with rusty-wooly undersides and down-rolled edges. It grows in boggy foothill to alpine sites from Alaska to southern BC and Alberta.

Glandular Labrador tea (*L. glandulosum*) has pale lower leaf surfaces, with dense resin glands among short, white scales. It grows in moist montane and subalpine sites from BC and Alberta to Wyoming.

WARNING:
Large amounts of Labrador tea can cause drowsiness, increased urination and intestinal disturbance. Both species contain narcotic compounds and toxins, which can cause cramps, delirium, palpitations, paralysis and even death. Glandular Labrador tea is poisonous to livestock, especially sheep. Some sources suggest boiling for long periods to destroy the alkaloids, but others say that boiling releases ledol, so it is not recommended. Do not confuse Labrador tea with the poisonous bog-laurels (p. 237). Pregnant women should not use Labrador tea.

Pipsissewa

Chimaphila umbellata

ALSO CALLED: prince's-pine.

FOOD: Pipsissewa has been used to flavor candy, soft drinks (especially root beer) and traditional beers. Unfortunately, this little plant is a secret ingredient in some commercial soft drinks and is disappearing from parts of the northwestern US at an alarming rate as a result of intensive collecting. The stems and roots were used to make tea, and the leaves and berries have been used as a trail nibble, but they are too tough and astringent to be eaten in large quantities.

MEDICINE: Native peoples ate the leaves raw or boiled the roots to make tonics rich in vitamin C. Pipsissewa tea was used as a remedy for fluid retention, kidney and bladder problems, fevers, colds, sore throats, coughs, backaches and stomachaches. It was also used as eye drops to relieve irritation from heat, smoke or perspiration. Pipsissewa has been shown to have astringent properties, which could reduce secretions in the eyes. This astringency, together with its reported anti-bacterial properties and ability to increase urine flow, would be useful for treating kidney infections. The tea was also applied externally to heal sores and rashes. The fresh, bruised leaves were applied to the skin to cause redness and blistering, as part of the treatment for heart and kidney diseases, tuberculosis of the lymph nodes and chronic rheumatism. Interestingly, a solution made by soaking the leaves in warm water was applied to heal blisters. The berries were eaten as a digestive tonic.

OTHER USES: Some tribes dried and ground the leaves and used them for smoking.

DESCRIPTION: Semi-woody evergreen shrub, 4–12" (10–30 cm) tall, with whorls of 3–8 narrowly spoon-shaped, glossy green leaves ³/₄–3" (2–8 cm) long. Flowers light pink or rose-tinged, waxy, about ³/₈" (1 cm) across, nodding above the leaves in erect clusters of 3–8, in June to August. Fruits round capsules, about ¹/₄" (5–7 mm) wide. Pipsissewa grows in foothill and montane woods (usually coniferous) from BC and Alberta to Colorado.

WARNING:
Pipsissewa has shown hypoglycemic activity in experiments. Leaves, bruised and held against the skin, can cause reddening, blistering and peeling.

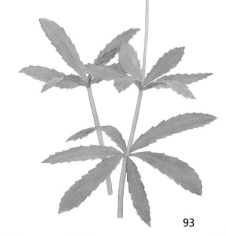

False-wintergreens

Gaultheria spp.

FOOD: The young leaves were occasionally eaten raw, but usually they were used to make a pleasant tea with a mild wintergreen flavor. The small, sweet berries have been eaten fresh, served with cream and sugar or cooked in sauces. They are said to make excellent jams and preserves (though it is necessary to add pectin). The berries were added to teas and were used to add fragrance and flavor to liqueurs. Occasionally, large quantities were picked and dried for winter use.

MEDICINE: All false-wintergreens contain methyl salicylate, a close relative of aspirin, which has been used to relieve aches and pains. These plants were widely used in the treatment of painful, inflamed joints, resulting from rheumatism and arthritis. The berries were soaked in brandy, and the resulting extract was taken to stimulate appetite, as a substitute for bitters. Studies suggest that oil-of-wintergreen is probably an effective pain-killer.

Slender false-wintergreen (top)
Hairy false-wintergreen (above)

OTHER USES: The leaves, plants and berries were boiled in water to add to baths for pregnant women.

DESCRIPTION: Delicate creeping shrublets with leathery, evergreen leaves, small, white to greenish or pinkish flowers, mealy to pulpy, fleshy, berry-like capsules and a mild wintergreen flavor.

Hairy false-wintergreen (*G. hispidula*) has stiff, flat-lying, brown hairs on its stems and lower leaf surfaces. Its leaves are very small (1/8–3/8" [4–10 mm] long), and it has tiny (1/16" [2 mm]), 4-lobed flowers and small (1/4" [5–7 mm]), white berries.

Alpine false-wintergreen (*G. humifusa*) has 1/2–3/4" (1–2 cm) long, glossy leaves, pinkish or greenish-white, 5-lobed, 1/8" (3–4 mm) wide flowers and scarlet, pulpy berries 1/4" (5–6 mm) wide. It grows in moist to wet, subalpine to alpine sites from BC and Alberta to Colorado.

Slender false-wintergreen (*G. ovatifolia*) is very similar to alpine wintergreen, but it has hairy (rather than hairless) calyxes and 3/4–2" (2–5 cm) long leaves. It grows on dry (occasionally boggy) subalpine sites in BC, Idaho and Montana.

WARNING:
Oil-of-wintergreen (most concentrated in the berries and young leaves) contains methyl salicylate, a drug that has caused accidental poisonings. It should never be taken internally, except in very small amounts. Avoid applying the oil when you are hot, because dangerous amounts could be absorbed through your skin. It can cause skin reactions and severe (anaphylactic) allergic reactions. People who are allergic to aspirin should not use false-wintergreen or its relatives.

Black huckleberry

Vaccinium membranaceum

ALSO CALLED: mountain huckleberry
• *V. globulare.*

FOOD: Some people consider this berry the most delicious and highly prized in the Rockies, while others think blueberries are best. Huckleberries can be used like domestic blueberries—eaten fresh from the bush, added to fruit salads, cooked in pies and cobblers, made into jams and jellies, or crushed and used in cold drinks. They are also delicious in pancakes, muffins, cakes and puddings. Huckleberries are often collected in large quantities in open, subalpine sites (such as old burns), and in some regions they are sold commercially. Native peoples gathered them in July to September and ate them fresh, but they also sun-dried or smoke-dried many for winter use. The berries were mashed and formed into cakes or spread loosely on mats for drying. Later, they were reconstituted by boiling either alone or with roots. Dried huckleberry leaves and berries are said to make excellent teas.

MEDICINE: The leaves and berries are high in vitamin C. Roots and stems of black huckleberry were used to make medicinal teas for treating heart trouble, arthritis and rheumatism. The leaf tea has been taken by some diabetics to stabilize and reduce blood sugar levels (reducing the need for insulin) and to treat hypoglycemia. Huckleberry-leaf tea has also been used as an appetite stimulant and as a treatment for bladder and urinary tract infections (see discussion, p. 98).

DESCRIPTION: Deciduous shrubs, 1–5' (0.3–1.5 m) tall, with thin, bright green, ³/₄–2" (2–5 cm) long, finely toothed leaves that turn red or purple in autumn. Flowers creamy pink to yellowish-pink, round to urn-shaped, ¹/₄" (5–6 mm) long, nodding on single, slender stalks, in April to June. Fruits black to dark purple berries, ³/₈" (8–10 mm) across. Black huckleberry grows on moist, open or wooded slopes in foothills and montane zones from the southern Yukon and NWT to Wyoming.

WARNING:
Large quantities of these berries can cause diarrhea.

95

Blueberries

Vaccinium spp.

FOOD: These sweet, juicy berries can be eaten fresh from the bush, added to fruit salads, cooked in pies, tarts and cobblers, made into jams, syrups and jellies, or crushed and used to make juice, wine, tea and cold drinks. Bog blueberry wine was said to be especially intoxicating. Blueberries also make a delicious addition to pancakes, muffins, cakes and puddings. They were widely used by native peoples, either fresh or dried singly or in cakes. To make cakes, the berries were boiled to a mush, spread in slabs, and dried on a rack in the sun or near a fire. The juice was slowly poured onto the drying cakes or was cooled to make jelly. Because blueberries grow close to the ground, they are difficult to collect, so some people combed them from their branches with a salmon backbone or wooden comb.

MEDICINE: These delicious berries are rich in vitamin C. Blueberry roots were boiled to make medicinal teas that were taken to relieve diarrhea, gargled to soothe sore mouths and throats, or applied to slow-healing sores. Bruised roots and berries were steeped in gin, which was to be taken freely (as much as the stomach and head could tolerate) to stimulate urination and relieve kidney stones and water retention. Blueberry-leaf tea and dried blueberries have been used like cranberries (see discussion on p. 98) to treat diarrhea and urinary tract infections. Low bilberry contains anthocyanosides, which are said to improve night-vision. These compounds are most concentrated in the dried fruit, preserves, jams and jellies. Their effect is said to wear off after 5–6 hours. Anthocyanins may reduce leakage in small blood vessels (capillaries), and blueberries have been suggested as a safe and effective treatment for water retention (during pregnancy), hemorrhoids, varicose veins and similar problems. They have also been recommended to reduce inflammation from acne and other skin problems and to prevent cataracts. Blueberry-leaf tea has been used by people suffering from hypoglycemia and by some diabetics, to stabilize and reduce blood sugar levels, and to reduce the need for insulin. The leaf or root tea of low bilberry is reported to flush pinworms from the body.

Oval-leaved blueberry (both photos)

OTHER USES: Blueberry leaves were sometimes dried and smoked. The berries were used to dye clothing navy blue.

DESCRIPTION: Low, often matted shrubs with thin, oval leaves 1–3 cm long, whitish to pink, nodding, urn-shaped flowers ¹/₈–¹/₄" (4–6 mm) long and round, ¹/₄–³/₈" (5–8 mm) wide, bluish berries, usually with a grayish bloom.

Low bilberry (*V. myrtillus*) has green, angled branches and finely toothed leaves and produces dark bluish-black to dark red fruits without a bloom in August to October. It grows on wooded montane and subalpine slopes from BC and Alberta to New Mexico.

Oval-leaved blueberry (*V. ovalifolium*) also has green, angled branches, but its leaves are blunt and toothless. It produces purple berries with a whitish bloom in early July to September in subalpine forests from BC and Alberta to Montana.

Dwarf blueberry (*V. cespitosum*) has rounded, yellowish to reddish branches and finely toothed, light green leaves. Its 5-lobed flowers produce single blueberries in August to September on montane, subalpine and alpine slopes from the Yukon and NWT to Colorado.

Bog blueberry (*V. uliginosum*) also has rounded, brownish branches, but it has toothless, dull green leaves and clustered (1–4), 4-lobed flowers. It grows in bogs and on alpine slopes from Alaska to Alberta and northern BC.

Bog blueberry (both photos)

WARNING:
Blueberry leaves contain moderately high concentrations of tannins, so they should not be used continually for extended periods of time.

97

Cranberries

Vaccinium spp.

FOOD: Cranberries can be very tart, but they make a refreshing trail nibble, and they are also added to fruit salads or cooked in pies and cobblers (usually mixed with other fruits). They make delicious jams and jellies, and they can be crushed or chopped to make tea, juice or wine and to flavor cold drinks. Cranberry sauce is still a favorite with meat or fowl. It is easily made by boiling berries with sugar and water or more traditionally by mixing cranberries with maple sugar and cider. Cranberries are a delicious addition to pancakes, muffins, breads, cakes and puddings. Firm, washed berries keep for several months, when stored in a cool place. They can also be frozen or canned. Native peoples sometimes dried cranberries for use in pemmican, soups, sauces and stews. Some tribes stored boiled cranberries mixed with oil and later whipped this mixture with snow and fish oil to make a dessert. Freezing makes cranberries sweeter, so they were traditionally collected after the first heavy frost. Because they remain on the shrubs all year, cranberries can be a valuable survival food. These low-growing berries are difficult to collect, so some people combed them from their branches with a salmon backbone or wooden comb.

Grouseberry (top); Bog cranberry (above)

MEDICINE: Cranberry juice has long been used to treat bladder infections. Research shows that these berries contain arbutin, which prevents some bacteria from adhering to the walls of the bladder and urinary tract and causing an infection. Cranberry juice also increases the acidity of the urine, thereby inhibiting bacterial activity, which can relieve infections. Increased acidity also lessens the urinary odor of people suffering from incontinence. Cranberries were also taken to relieve nausea, to ease cramps and childbirth and to quiet hysteria and convulsions. Crushed cranberries were used as poultices on wounds, including poison-arrow wounds.

OTHER USES: The red pulp (left after the berries have been crushed to make juice) contains red dyes. Hunters often looked for this plant when seeking grouse, because it is a favorite food of these game birds.

WARNING:
Large quantities of cranberries can cause diarrhea.

DESCRIPTION: Dwarf, deciduous shrubs, mostly less than 4" (10 cm) tall, with small, nodding, pinkish flowers producing sour, bright red (sometimes purplish) cranberries.

Bog cranberry (*V. oxycoccos*, also called *Oxycoccus quadripetalus* and *O. microcarpus*) has slender, creeping stems, with small (mostly less than ³/₈" [1 cm] long), pointed, glossy leaves. Its distinctive little shooting star–like flowers produce deep red cranberries about ¹/₄–³/₈" (5–10 mm) wide in July to August. Bog cranberry grows in bogs from Alaska to Idaho.

Grouseberry or **whortleberry** (*V. scoparium*) is a low, broom-like shrub with many slender, green, angled branches. It has thin, finely toothed leaves ¹/₄–¹/₂" (6–12 mm) long and urn-shaped flowers. Its small (¹/₈–¹/₄" [3–5 mm]), single, bright red berries appear in July to August on foothill, montane and subalpine slopes from BC and Alberta to Colorado.

Lingonberry or **mountain cranberry** (*V. vitis-idaea*) has ¹/₄–⁵/₈" (6–15 mm) long, blunt, leathery, evergreen leaves, with dark dots (hairs) on a pale lower surface. Its clusters of urn-shaped flowers produce bright red cranberries in July to August. Lingonberry grows on foothill, montane, subalpine and alpine slopes from Alaska to BC and Alberta.

Grouseberry

Lingonberry

Devil's club

Oplopanax horridus

FOOD: Roots and young, fleshy stems were cooked and eaten (see Warning). Young leaves, collected before the spines stiffen, may be eaten raw or cooked in soups and casseroles.

MEDICINE: The root tea has been reported to stimulate the respiratory tract and to help bring up phlegm when treating colds, bronchitis and pneumonia. It has also been used to treat diabetes, helping to regulate blood sugar levels and to reduce the craving for sugar. Although this tea is recommended for binge-eaters who are trying to lose weight, some tribes used it to improve appetite and to help people gain weight. In fact, it was said that a patient could gain too much weight if it was used for too long. Devil's club extracts have lowered blood sugar levels in laboratory animals. Crushed, fresh stems were steeped in hot water to make teas for relieving indigestion and constipation, or they were boiled to make a blood purifier, tonic and laxative. A strong decoction induced vomiting in purifying rituals preceding important events such as hunting or war expeditions. This decoction can also be applied to wounds to combat staphylococcus infections, and ashes from burned stems were sometimes mixed with grease to make salves for healing swellings and weeping sores. Like many members of the ginseng family, devil's club contains glycosides that are said to reduce metabolic stress and thus improve one's sense of well-being.

OTHER USES: The aromatic roots were smoked with tobacco (*Nicotiana tabacum*), and the dried, powdered bark was ground to make sweet-smelling baby powder and deodorant. Devil's club berries were rubbed into hair to combat dandruff and lice and to add shine.

DESCRIPTION: Strong-smelling, deciduous shrub with spiny, erect or sprawling stems $3^1/_2$–10' (1–3 m) tall. Leaves broadly maple leaf–like, 4–16" (10–40 cm) wide, with prickly ribs and long, bristly stalks. Flowers greenish-white, $^1/_4$" (5–6 mm) long, 5-petaled, forming 4–10" (10–25 cm) long, pyramid-shaped clusters, in May to July. Fruits bright red, berry-like drupes $^1/_8$–$^1/_4$" (4–6 mm) long, in showy clusters. Devil's club grows in moist, shady foothill and montane sites from BC and Alberta to Idaho and Montana.

WARNING:
The berries are inedible. The spines break off easily and often cause infections. Some people have an allergic reaction to their scratches.

Fool's onion

Triteleia grandiflora

ALSO CALLED: white hyacinth
• gophernuts • *Brodiaea douglasii.*

FOOD: Fool's onion was one of the first roots to be dug in spring. It was collected in large quantities by both native peoples and settlers, often with the bulbs of yellowbells (p. 109). Plants had to be marked the previous year, so that their corms could be located the following spring, before shoots appeared above ground. Fool's onion was sometimes eaten fresh, but it was said to be rather gluey when raw. The flavor improved with a few minutes of simmering, but the corms were considered best roasted in hot ashes or over a slow fire for about an hour. Cooking breaks down some of the starches and makes the corms much sweeter. Some tribes dried the boiled corms for later use. The young seed pods were also eaten raw, as a nibble, or cooked as a green vegetable.

MEDICINE: Fool's onion was reported to have saved at least 1 family from starvation. It was believed to have some magical powers and was included in medicine bags to increase the potency of the bag.

DESCRIPTION: Slender, perennial herb, 8–24" (20–60 cm) tall, with 1–2 flat, grass-like leaves above a scaly, bulbous stem base (corm) about $3/4$" (2 cm) thick. Flowers deep to light blue, about 2 cm long and wide, bell-shaped, with 3 ruffled inner petals and 3 smooth-edged outer petals, forming flat-topped clusters (umbels), in April to July. Fruits rounded capsules, $1/4$–$3/8$" (6–10 mm) long. Fool's onion grows on dry, open or wooded slopes in plains, foothills and montane zones from southern BC to Wyoming.

WARNING:
These corms could be confused with the bulbs of death-camases (p. 240). Never eat wild roots or bulbs unless you are absolutely sure of the identity of the plant.

Wild onions
Allium spp.

Nodding onion

FOOD: All of the *Allium* species are edible, though some are much stronger than domestic onions. These plants were widely eaten by native peoples and European settlers, either raw or cooked. Wild onions can be served as a hot vegetable in cream sauce, but usually they were used to flavor other greens and roots or were fried or boiled with meat in casseroles, soups and stews. Cooking removes the strong smell and flavor, converting the sugar inulin to the more digestible fructose, and the bulbs become very sweet. The tender young leaves of most onions (before flowering) have been used raw in sandwiches and salads or cooked with the bulbs. The bulbs were sometimes rubbed in hot ashes to singe off the outer fibers and remove some of the strong taste. They were then eaten immediately or dried and stored for winter. Some tribes layered onions between alder or saskatoon branches in cooking pits and steamed them for several hours, until they were sweet and almost black. Often, onions were steamed with camas bulbs (p. 105) or hair lichen (pp. 232–33). The sweet, cooked bulbs were usually eaten immediately, but they could also be dried, singly or in cakes, for future use.

MEDICINE: Onions are reported to have anti-bacterial, anti-viral and anti-fungal qualities, and they have been used for many years in the treatment of cuts, burns and insect bites and stings. The juice from the bulbs was boiled to make a syrup, sometimes sweetened with honey, for treating colds and sore throats. The bulbs have traditionally been used to relieve indigestion, gas and vomiting and were reputed to cure sexual impotency caused by mental stress or illness. They were also dried, ground and used as snuff for opening the sinuses.

WARNING:
Wild onions resemble and often grow in the same habitats as their poisonous relative, mountain death-camas (p. 240). Some tribes believed that wild onions in the mountains were poisonous, probably because of earlier confusion with death-camas. Mountain death-camas does not smell like an onion. If it doesn't smell like an onion, don't eat it. Some people develop skin reactions after handling onions.

Frequent doses of onions have been shown to reduce blood lipid (cholesterol) levels and blood pressure and to prevent blood clots from forming.

OTHER USES: Wild onions make an attractive, hardy addition to wildflower gardens and have the added benefit of being edible.

DESCRIPTION: Slender, perennial herbs with oval bulbs, smelling strongly of onions. Flowers bell-shaped, 6-tepaled, clustered at the tips of slender, leafless stems.

Nodding onion (*A. cernuum*) has bulbs coated with parallel (not netted) fibers, pinkish-mauve to white flowers in nodding clusters, and capsules tipped with 6 small crests. It grows on open plains, foothill and montane sites from BC and Alberta to New Mexico.

Short-styled onion (*A. brevistylum*) is similar, but it has erect flowers and crestless capsules. It grows in wet, montane and subalpine sites from Idaho and Montana to Colorado.

Two common onions have erect flowers and netted, fibrous bulb coats. **Geyer's onion** (*A. geyeri*) has 3 or more leaves and (usually) deep pink flowers, whereas **prairie onion** (*A. textile*) usually has 2 leaves and white flowers. Both species grow on plains, foothill and montane slopes from Alberta to New Mexico.

Nodding onion (all images)

Wild chives

Allium schoenoprasum

FOOD: Wild chives has been used like domestic varieties for flavoring soups, salads, sandwiches, vegetables and meat dishes. The leaves, bulbs and flowerheads are all edible. Wild chives can be strong by itself, but it has been pickled or used as a cooked vegetable, served hot with butter. Chives contain less sulfur than most onions and are more easily digested. Chives were dried or frozen for future use or packed with alternate layers of rock salt and stored in a cool place.

MEDICINE: The most common medicinal use of chives was in treating coughs and colds. The juice was either boiled down to a thick syrup or a sliced bulb was placed in sugar and the resulting syrup was taken. Dried chive bulbs were burned in smudges to fumigate patients, or they were ground and inhaled like snuff to clear the sinuses. Wild chives was also said to stimulate appetite and aid digestion. On the other hand, water in which it had been crushed and soaked for 12 hours was swallowed on an empty stomach to rid the system of worms. Crushed plants were used to treat insect bites and stings, hives, burns, scalds, sores, blemishes and even snakebites.

OTHER USES: The flower clusters dry well and make a beautiful purplish addition to dried flower arrangements. Chives is an attractive, edible garden plant that flowers throughout the growing season. It is easily propagated from seed or by dividing. Some gardeners have recommended chives as a companion plant for carrots, grapes, roses and tomatoes. It has also been reported to deter Japanese beetles and black spot on roses, scabs on apples and mildew on cucumbers. Chive leaves and bulbs were sometimes rubbed onto skin and clothing as an insect repellent.

DESCRIPTION: Slender, perennial herb with hollow, cylindrical leaves from elongated bulbs about ³/₈" (1 cm) thick, smelling strongly of onions. Flowers deep pink to lilac or white, bell-shaped, 6-tepaled, forming clusters at the tips of hollow, leafless stems, in April to August. Fruits small, membranous capsules, containing shiny, black seeds. Wild chives grows in moist, open sites in plains, foothills and montane zones from Alaska to Colorado.

WARNING:
Without flowers, wild chives might be confused with its poisonous relatives, the death-camases (p. 240), but those plants have flat (not hollow) leaves and no onion-like smell.

Common camas

Camassia quamash

ALSO CALLED: blue camas.

FOOD: Camas provided one of the most prized root crops, and some tribes fought for the right to collect in certain meadows. Its role has been likened to that of cereal plants in Europe. Girls competed in the annual camas harvest to show their worth as future wives. One young woman was reported to have collected and prepared 60, 9-gal (42-ℓ) sacks of camas roots. Men were excluded from the harvest, because they would bring bad luck, and if a woman burned any bulbs, it was said that some of her relatives would soon die. The bulbs were used singly or pounded and formed into cakes. They were eaten raw, roasted or stone-boiled, but most were cooked and dried. Usually, bulbs were baked in pits with hot stones for several days until they were dark brown, with a glue-like consistency and a sweet taste, like that of molasses. They were then mashed together and made into cakes, which were sun-dried for storage. During the cooking process, inulin (an indigestible sugar) breaks down to fructose, which is sweet and easily digested. Some cooked dried bulbs are 43 percent fructose by weight. Camas was the principal sweetening agent for many tribes prior to the introduction of sugar. Cooked dried bulbs

were sometimes ground and made into small cakes that were eaten with flour, cream and sugar. Concentrated liquid from boiling the bulbs was made into a sweet, hot drink or was mixed with flour to make gravy. In 1847, David Thompson reported eating 36-year-old bulbs that still had good flavor, though they had probably lost their aroma after 2 years.

WARNING:
Camas can cause gas and indigestion in the uninitiated; large quantities can result in vomiting and diarrhea. These bulbs could be confused with those of the death-camases (p. 240). It is safest to collect them when the plants are in flower. Camas is easily identified by its showy, blue flowers.

DESCRIPTION: Slender, perennial herb, 8–24" (20–60 cm) tall, with oval bulbs $^{1}/_{2}$–1$^{1}/_{4}$" (1–3 cm) thick and grass-like leaves. Flowers pale to deep blue or violet, about 1$^{5}/_{8}$–2" (4–5 cm) across, with 6 linear, 3–9-nerved petals, forming 2–12" (5–30 cm) long clusters, in May to June. Fruits oval capsules, $^{1}/_{2}$–$^{5}/_{8}$" (12–15 mm) long. Common camas grows in moist meadows in plains and foothills zones from southern BC and southwestern Alberta to Wyoming.

Beargrass
Xerophyllum tenax

FOOD: The extensive, stringy roots are edible, but it was recommended that they be roasted or boiled.

OTHER USES: This large, common plant provides an abundant supply of fiber. The tough, fibrous leaves were pounded to separate their fibers, which were then twisted together to make ropes and cord. Some tribes wove the long, slender leaves into hats, baskets and capes. Beargrass is difficult to grow, and it does not do well in gardens, so it is best left in its natural habitat. Although beargrass is very showy when in bloom, most plants flower only once every 3–10 years. Often, all of the plants in a population bloom together, covering a slope with white and filling the air with a fragrant, lily-like perfume.

DESCRIPTION: Robust, evergreen, perennial with large clumps of tough, wiry, grass-like leaves 8–24" (20–60 cm) long and edged with sharp, fine teeth. Flowers fragrant, white, about $5/8$" (1.5 cm) across, with 6 spreading petals, forming showy, club-shaped, bottle-brush-like clusters $1^1/_2$–5' (50–150 cm) tall, in May to August. Fruits dry, oval, 3-lobed capsules $1/4$" (5–7 mm) long. Beargrass grows on dry, open (occasionally forested) slopes in montane, subalpine and alpine zones from southeastern BC and southwestern Alberta to Idaho and Montana.

Narrow-leaved yucca

Yucca glauca

ALSO CALLED: soapwort.

FOOD: The flowers have been eaten raw in salads, cooked in soups, or deep-fried like squash blossoms. It is recommended that their bitter, green centers be removed. The young seed pods were roasted in ashes and eaten, and the ripe fruits were peeled, seeded and baked. They have been used like apples in pies and jams. The leaves are also said to have been boiled with salt and eaten, but they are extremely fibrous.

MEDICINE: Tea made from the roots was given to women in prolonged labor. It was thought to clean the sticky covering from an over-sized baby and speed delivery. Similarly, a cupful of yucca suds sweetened with sugar was given to speed delivery of the afterbirth. Yucca fruit has been reported to induce vomiting, but most sources say that it is edible.

OTHER USES: The fibrous leaves were split and used as all-purpose ties, for hanging and drying household items and for weaving loose baskets. Some tribes pounded the leaves in water to release the fibers and then twisted or plaited them to make cord, rope and nets. Narrow slips of leaves were fringed and used as paintbrushes. About 80 million pounds (36 million kilograms) of yucca fiber was harvested in Texas and New Mexico during World War I and used to make bags and burlap. The Navajo pounded the roots with rocks to remove bark and soften the roots and then vigorously stirred the softened mass of fibers in warm water to whip up suds. This soapy mixture was used to wash wool, clothing, hair and the body. It was also said to reduce dandruff and baldness. The Ramah Navaho used the soapy mixture for ceremonial purification baths before weddings and the naming of infants and after burials. The juice of the leaves, boiled alone, was used as a red dye. Yucca is the state flower of New Mexico.

DESCRIPTION: Coarse, evergreen perennial with dense tufts of stiff, linear, sharp-pointed leaves 8–24" (20–60 cm) long, edged with whitish, frayed fibers. Flowers cream-colored to greenish-white, $1^5/_8$–2" (4–5 cm) long and wide, with 6 leathery, oval petals, nodding, in showy, $1^1/_2$–5' (50–150 cm) tall clusters, in May to July. Fruits hardened, oblong capsules 2–$2^3/_4$" (5–7 cm) long. Narrow-leaved yucca grows on dry, open slopes in plains and foothills zones from southern Alberta to New Mexico.

Mariposa-lilies
Calochortus spp.

FOOD: The bulbs of mariposa-lilies are crisp, sweet and nutritious, and they were eaten by many native peoples and settlers. They were sometimes used raw in salads, but usually they were boiled, roasted in ashes or over a smoky fire, or steamed in fire pits. Some bulbs were dried for future use. Dried bulbs were boiled in soups and stews or ground into flour for thickening soups and supplementing breads. More recently, these bulbs have been preserved by canning. In 1848, sego-lily bulbs were reported to have saved the lives of many Mormons, when crops were destroyed by crickets, drought and frost. The nectar-rich flower buds of some species were eaten raw, as a sweet nibble.

MEDICINE: Juice from the leaves was applied to pimples, and the whole plant was boiled to make a medicinal tea that was given to women in labor to facilitate delivery of the afterbirth.

OTHER USES: Sego-lily is the state flower of Utah.

DESCRIPTION: Slender, perennial herbs with deep, fleshy, onion-like bulbs. Leaves grass-like and mainly basal. Flowers broadly cupped, about $3/4$–$1^5/8$" (2–4 cm) across, with 3 wide, rounded or abruptly pointed petals, each with a distinctive gland, band of color and/or fringe of hairs near the base, borne in loose clusters of 1–5.

Three-spotted mariposa-lily (*C. apiculatus*) has flat leaves, yellowish-white flowers, with 3 small, round, purplish-black dots, and nodding, 3-winged capsules. It grows on foothill and montane slopes in BC, Alberta, Idaho and Montana.

Sego-lily (*C. nuttallii*) has ivory-white flowers with 3 large, dark glands, each below a crescent-shaped, reddish-brown or purple band. It grows on dry foothill and montane slopes from Idaho and Montana to New Mexico.

WARNING: Harvesting the bulb destroys the plant. Collecting and overgrazing by cattle and sheep have resulted in the loss of many populations of this beautiful wildflower. Mariposa-lilies should be eaten only in an emergency.

Three-spotted mariposa-lily (all images)

Fritillaries

Fritillaria spp.

FOOD: These fleshy bulbs produce tiny bulblets about the size of rice grains, which have been called 'root rice.' Both bulbs and bulblets are edible, either raw or cooked. Several tribes gathered these roots in early spring and boiled them alone or with bitterroot (p. 118), which was collected at the same time. They were also steamed in pits or dried for winter use. The fruiting pods were occasionally boiled as a wild green, but they were not used by most tribes.

MEDICINE: Leopard lily bulbs were pulverized and used to make salves for treating tuberculosis of the lymph nodes in the neck (scrofula). This disease was associated with 'gophers' (ground squirrels), and children were warned to keep away from 'gopher' mounds or they would get swollen necks, just like the cheek pouches on the animals.

OTHER USES: Several fritillaries are popular garden flowers. Perhaps the best known is 'mission bells.' Some people find the smell of the dark-flowered species very unpleasant, but it is attractive to flies, which probably pollinate the flowers during their visits.

Yellowbells (both photos)

DESCRIPTION: Slender, perennial herbs, from scaly bulbs surrounded by tiny, rice-like bulblets. Leaves alternate to whorled, slender, rather fleshy. Flowers bell-shaped, nodding, about $1/2–3/4$" (1–2 cm) long, with 6 petals, single or in loose clusters of 2–4. Fruits small capsules.

WARNING:
Many members of the Lily family appear very similar before they flower, and several are poisonous. Some tribes considered yellowbells poisonous, probably because of confusion with other plants. Only plants with flowers or flower remains could be positively identified. Harvesting the bulb destroys the plant, and over-collecting can eradicate populations of this beautiful wildflower. These plants should be used only in an emergency.

Yellowbells (*F. pudica*) is a 4–12" (10–30 cm) tall plant, with yellow, usually single flowers that fade to reddish or purplish. It grows on moderately dry plains, foothill and montane slopes from BC and Alberta to Colorado.

Leopard lily or **purple fritillary** (*F. atropurpurea*) is a 6–24" (15–60 cm) tall plant, with 1–4 nodding, brown flowers marked with yellow or white blotches. It grows on grassy plains, foothill and montane slopes from southern Idaho to New Mexico.

Lilies

Lilium spp.

Wood lily (top); Tiger lily (above)

FOOD: The flowers, seeds and bulbs have all been used as food. Although the bulbs have a strong, bitter, peppery flavor, they were very popular among native peoples. They were generally cooked and eaten with other foods, such as venison, fish and saskatoons, as a flavoring and thickening agent for stews. Some tribes also ate lily bulbs fresh or dried them for winter storage. Usually, lily bulbs were boiled in 2 changes of water or steamed in fire pits. Cooked bulbs were also dried for winter use, either singly or mashed and formed into thin cakes. The flowers have been used in salads and are said to be delicious as well as beautiful.

MEDICINE: Northern native peoples used wood lily roots to make medicinal teas that were taken to treat stomach disorders, coughs, tuberculosis and fevers and to help women in labor deliver the afterbirth. These teas were also used as a wash for swellings, bruises, wounds and sores. The flowers and the root tea were used in poultices for treating spider bites.

OTHER USES: Lilies make beautiful garden flowers and are easily transplanted but see Warning.

DESCRIPTION: Hairless, perennial herbs with single, erect stems from whitish, scaly bulbs. Leaves alternate to whorled, 2–4" (5–10 cm) long. Flowers showy, usually dark-dotted in the throat, appearing in June to August. Fruits erect, cylindrical capsules, 3/4–15/8" (2–4 cm) long.

Wood lily (*L. philadelphicum*) has whorled upper leaves (usually in 6s) and 1–3 large (about 3" [8 cm] wide), erect, goblet-shaped flowers, with orange to brick-red petals and paler, purplish-dotted throats. It grows on moist plains, foothill and montane slopes from BC and Alberta to New Mexico.

Tiger lily or **Columbia lily** (*L. columbianum*) has 2–20 smaller (1 1/4–1 5/8" [3–4 cm] wide), nodding, orange flowers, with purple-spotted petals that curl back towards the base. It grows on foothill, montane and subalpine slopes in BC and Idaho.

WARNING:
Lilies are seldom plentiful, and when they are picked or mowed, all of the leaves are removed with the stalk and the plant dies. Digging, mowing and over-picking have resulted in the near extinction of this beautiful wildflower in many populated areas. Please leave them for others to enjoy.

Yellow glacier-lily
Erythronium grandiflorum

ALSO CALLED: snow-lily.

FOOD: These slender bulbs were collected in large quantities by many tribes, and strings of dried bulbs were a popular trade item. The bulbs can be eaten raw, but like many bulbs and roots, they are made sweeter and more easily digestible by long, slow cooking. Drying also helps this process. Usually they were steamed, roasted or boiled, and many were dried for winter use. A hundred kilograms or more was considered a good winter supply for a family. Dried bulbs were soaked and then boiled or steamed. Once cooked, the bulbs became chocolate-brown, soft and sweet. Most tribes used only the bulbs, but the leaves were also eaten occasionally, raw or cooked. Also, the fresh green seed pods were said to taste like string beans when cooked.

MEDICINE: The leaf tea has been shown to kill a wide range of bacteria and can be used as an antiseptic wash for cuts, scrapes and sores. Compounds extracted from these plants have also been shown to be slightly anti-mutagenic and to have tumor-reducing qualities. Teas made from other species of glacier-lily have been used to treat fevers, swelling and infection and to reduce the chances of conception.

OTHER USES: This beautiful wildflower would make a lovely addition to a wildflower garden, but it can be difficult to cultivate.

DESCRIPTION: Perennial herb with 2 leaves (4–8" [10–20 cm] long) from deeply buried, elongated, corm-like bulbs. Flowers bright yellow, nodding, $1^1/4$–$2^1/2$" (3–6 cm) across, with 6 petals curved upwards and 6 large stamens projecting downwards, usually solitary on leaf-less, 4–16" (10–40 cm) tall stalks, in April to August. Fruits erect, 3-sided, club-shaped capsules, $1^1/4$–$1^5/8$" (3–4 cm) long. Yellow glacier-lily grows on moist, rich, shaded to open slopes in montane, subalpine and alpine zones from BC and Alberta to Colorado and Utah.

WARNING:
The bulbs sometimes cause a burning sensation, and too many can cause vomiting. Large quantities of the leaves and seed pods can cause vomiting and diarrhea. These small bulbs are best collected when in the plants are in bloom, to avoid confusion with poisonous members of the Lily family.

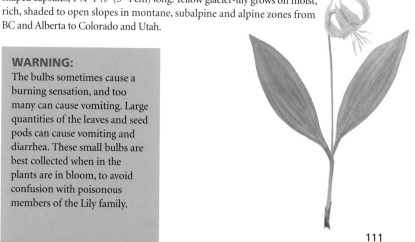

Mountain marsh-marigold

Caltha leptosepala

ALSO CALLED: elkslip.

FOOD: The broad, fleshy leaves have been boiled until tender (10–60 minutes depending on the plant) and served with butter or in cream sauce. Mountain marsh-marigold could be an important emergency food at high elevations, because it is often abundant in areas where other edible plants are scarce, but it would probably be wise to cook it if possible and to use it in moderation (see Warning). The fleshy roots have also been cooked and eaten. This dish is said to resemble sauerkraut.

DESCRIPTION: Fleshy, perennial herb with thick, waxy, oblong to heart-shaped leaves $^3/_4$–$2^1/_2$" (2–6 cm) long. Flowers white (sometimes bluish-tinged), about $^3/_4$–$1^5/_8$" (2–4 cm) across, with 5–15 oblong to oval, petal-like sepals and no petals, borne singly on leafless, 2–4" (5–10 cm) tall stems, in May to August. Fruits erect clusters of pods (follicles), about $^5/_8$" (15 mm) long. Mountain marsh-marigold grows on wet, open sites in upper montane, subalpine and alpine zones from BC and Alberta to Colorado.

WARNING:
Some sources warn against eating marsh-marigolds, but many writers recommend this plant as a potherb and some say they have used the raw flowers and young leaves in salads with no ill effects. Many marsh-marigolds contain volatile, poisonous glycosides that are driven off or destroyed by boiling or drying.

Rocky Mountain cow-lily

Nuphar lutea

ALSO CALLED: yellow pond-lily.

FOOD: The seeds of this aquatic plant were an important food for some tribes. Dried capsules were broken open and winnowed to separate the seeds, which were then popped and eaten like popcorn or fried in bear fat. Dried, fried or popped kernels were also ground into flour. Some people consider cow-lily rootstocks extremely bitter and unpleasant, even after prolonged boiling in several changes of water, but others have described them as sweet excellent eating. The rootstocks are probably best in late autumn or early spring. They were usually roasted or boiled and then peeled, sliced and eaten with meat in soups and stews. Thin, cooked slices were dried and then stored or ground into flour for making gruel and for thickening soups.

MEDICINE: Cow-lily rootstocks were used in medicinal teas to treat sore throats, inflamed gums, diarrhea, gallstones, stomach inflammation, sexual irritability, venereal disease, blood diseases, chills with fever, heart problems and impotence. Some present-day herbalists have recommended the seeds or dried rootstocks as a cooling astringent and anti-inflammatory for treating irritated digestive tracts, urinary tracts and reproductive organs. Mashed rootstocks have been used as poultices on wounds, bruises, boils, swellings and inflamed and swollen joints. They contain mucilage, tannin and steroids, and some of their alkaloids are reported to stimulate the heart, constrict blood vessels and relieve or prevent spasms. They also contain antagonistic alkaloids. For example, one reduces blood pressure and one increases blood pressure.

OTHER USES: Bruised rootstocks, steeped in milk, produced an insecticide for killing beetles and cockroaches. Smoke from burned rootstocks was said to drive away crickets.

DESCRIPTION: Aquatic, perennial herb with fleshy stems up to 6^1/$_2$' (2 m) long from massive rootstocks (to 16^1/$_2$' [5 m] long and 6" [15 cm] in diameter). Leaves floating, leathery, round to heart-shaped, 4–16" (10–40 cm) across. Flowers yellow, often tinged green or red, 2^1/$_2$–4" (6–10 cm) across, with about 6–9 waxy, petal-like sepals and with 10–20 tiny, true petals, partially hidden below the stamens and large, yellow, disc-like stigma, borne singly, on or above the water surface, in May to August. Fruits leathery, oval capsules. Rocky Mountain cow-lily grows in shallow, quiet water in plains to subalpine zones from Alaska to Colorado.

WARNING:
Large amounts of these rootstocks are potentially poisonous.

Bedstraws

Galium spp.

Sweet-scented bedstraw

FOOD: Bedstraws are related to coffee, and their nutlets have been dried, roasted until dark brown, ground, and used as a good caffeine-free coffee substitute. The dried leaves and roots are said to make a pleasant tea. The young, sweet-smelling plants are edible raw or cooked, but raw plants of rough-barbed species can cause gagging and are best cooked. The flavor of cooked bedstraw has been likened to that of spinach. Cooked, cooled shoots were sometimes used in salads.

MEDICINE: Bedstraws are rich in vitamin C and have been used in spring tonics and as a cure for scurvy. Of the 3 species described here, cleavers has been used most widely as a medicine. Plants were used in sweet-smelling hot compresses to stop bleeding and soothe sore muscles. They were also dried, ground and sprinkled onto cuts, scrapes and other wounds to stop bleeding and aid healing, or they were made into washes for treating freckles, sunburn, psoriasis, eczema, ulcers and other skin conditions. Crushed plants or fresh plant juice were used to stop nosebleeds and to soothe minor burns and skin irritations. Cleavers tea has been reported to stimulate urination and the drainage of lymph-engorged tissues and to accelerate the metabolism of stored fat. Consequently, it has been used as a treatment for kidney stones and bladder problems and as a weight-loss aid. It was traditionally used in medicinal teas and washes for treating cancer, and it has been found to contain citric acid (reported to have some anti-tumor activity) and asperuloside (an anti-inflammatory). Cleavers extracts have also been shown to lower blood pressure and to combat certain yeasts. Sweet-scented bedstraw contains substances that lower blood pressure.

OTHER USES: The sweet-smelling, vanilla-scented plants were used to stuff pillows and mattresses, hence the common name 'bedstraw.' They were also used as perfume. The roots of many species produce a red or purple dye. Bristly masses of cleavers make good temporary strainers.

DESCRIPTION: Leafy, perennial or annual herbs with whorled leaves on slender, 4-sided, erect to sprawling stems. Flowers small, white, 4-petaled, forming branched clusters, in June to August. Fruits hairy nutlet pairs, about ⅛" (2–3 mm) long.

WARNING:
People with kidney problems or poor circulation or with a tendency towards diabetes should not use these plants. Continual use of bedstraw will irritate the mouth, and the plant juice can cause rashes on sensitive skin.

Northern bedstraw (*G. boreale*) has whorls of 4 slender, blunt-tipped, 3-nerved leaves, and its nutlets have short, straight hairs. It grows in well-drained foothill, montane and subalpine sites from Alaska to New Mexico.

Sweet-scented bedstraw (*g. triflorum*) has whorls of 6 broader (over $1/8$" [3 mm] wide), bristle-tipped leaves, and its nutlets are covered with hooked bristles. It grows in moist plains, foothill and montane woods from the Yukon and NWT to New Mexico.

Cleavers (*G. aparine*) resembles sweet-scented bedstraw, but it is an annual species with whorls of 6–8 narrower, $1/8$" (2–3 mm) wide leaves. Cleavers is common on disturbed ground and moist, shady sites in the plains and foothills zones from BC and Alberta to New Mexico.

Northern bedstraw (all images)

115

Alpine fireweed

Fireweeds
Epilobium spp.

FOOD: Fireweeds have been widely used as greens, either raw or cooked. The young shoots have been likened to asparagus and the young leaves to spinach. The beautiful pink flowers make a colorful addition to salads, and flower-bud clusters can be cooked as a vegetable. Fireweed tea has been enjoyed around the world, and fireweed honey is popular in some regions. The stem pith was added to soups as a thickener or dried, boiled and fermented to make fireweed ale.

MEDICINE: Fireweed plants are rich in beta-carotene and vitamins A and C. Common fireweed has been most widely used as a medicine. Its leaf extracts have been shown to be anti-inflammatory, and it has been used to treat yeast infections (candidiasis), hemorrhoids, diarrhea, cramps and general inflammation in the digestive tract (mouth, stomach, intestine), either taken internally in teas or applied externally in washes, enemas and douches. Leaf and flower teas were sometimes used to treat asthma and whooping cough. Peeled roots were applied as poultices to burns, swellings, boils, sores and rashes, and the leaves were applied to mouth ulcers.

OTHER USES: Some tribes used its stem fibers to make cord and fish nets. Common fireweed is sometimes grown as an ornamental, but it can become a troublesome weed. The stem pith was dried, powdered and rubbed on the face and hands in winter to protect skin from the cold and to the reduce pain when warming cold hands. Fireweed flowers were sometimes rubbed into rawhide for waterproofing, and the fluffy 'down' in mature seed pods provided tinder for fires, padding for quilts and fibers for blankets and clothing.

DESCRIPTION: Clumped, perennial herbs with alternate, lance-shaped leaves on erect stems. Flowers pink to rose-purple (rarely white), 4-petaled, with a prominent, 4-pronged style, forming showy clusters, in June to September. Fruits erect, linear pods, splitting lengthwise to release 100s of fluffy-parachuted seeds.

Common fireweed (*E. angustifolium*) has spreading rootstocks with tall (1–10' [30–300 cm]) stems, producing clusters of 15 or more small ($^3/_4$–$1^5/_8$" [2–4 cm] wide) flowers. It grows on open, disturbed, foothill, montane and subalpine sites from Alaska to New Mexico.

Alpine fireweed or **broad-leaved fireweed** (*E. latifolium*) is a smaller (2–16" [5–40 cm] tall) plant without rootstocks and with clusters of 1–10 large ($^3/_4$–$2^1/_2$" [2–6 cm] wide) flowers. It grows on gravelly sites in montane, subalpine and alpine zones from Alaska to Colorado.

Common fireweed

WARNING: Some people find fireweed slightly laxative, so start with small quantities.

Common evening-primrose

Oenothera biennis

FOOD: The roots of young (first year) plants are said to taste like rutabagas or parsnips, though some people say they are an acquired taste. When gathered in late autumn or early spring, they were sometimes eaten raw, but usually they were boiled for 2 hours and/or in 2 changes of water to reduce their peppery flavor. Cooked roots have been served as a hot vegetable, fried, roasted with meat, sliced and added to soups and stews or boiled in syrup until they were candied. The young leaves, flower buds and green pods have also been cooked as greens, usually in 1–3 changes of water. Flowers and flower buds could make a colorful addition to salads. Evening-primrose roots and flower buds have also been suggested for pickling. The oil-rich seeds have been used like poppy seeds, sprinkled on breads and in salads.

MEDICINE: The roots or shoot tips were steeped in hot water or simmered in honey to make soothing teas and cough syrups, which were said to have antispasmodic, sedative effects. Some tribes used evening-primrose tea to treat obesity and laziness. Evening-primrose roots have been used in poultices on swellings, hemorrhoids and boils and rubbed on athletes' muscles to increase strength. The aromatic seeds are rich in essential fatty acids, such as gamma-linoleic acid and *cis*-linoleic acid, which the body converts into important hormones. Evening-primrose oil is said to reduce inflammation, reduce imbalances and abnormalities in prostaglandin production, and regulate liver functions. Clinical studies have shown evening-primrose oil to be useful for treating asthma, psoriasis, arthritis, weak immune systems, infertility, premenstrual syndrome and heart and vascular diseases. It has also been suggested as a remedy for eczema, rheumatoid arthritis, multiple sclerosis, migraine headaches, inflammations, menopausal and breast problems, diabetes, alcoholism (including liver damage) and cancer.

DESCRIPTION: Erect, biennial herb with single stems about $1^1/_2$–$3^1/_4$' (50–100 cm) tall from stout taproots. Leaves in basal rosettes (first year) and alternate (second year), $3/_4$–6" (2–15 cm) long. Flowers bright yellow, about $1^1/_4$–$1^5/_8$" (3–4 cm) across, with 4 broad petals and 4 backward-bending sepals at the tip of a $1^1/_4$–2" (3–5 cm) long calyx tube (hypanthium), forming long clusters, in June to August. Fruits erect, hairy, spindle-shaped capsules $3/_4$–$1^5/_8$" (2–4 cm) long. Common evening-primrose grows in moderately dry, open sites in plains, foothills and montane zones from BC and Alberta to New Mexico.

Bitterroot
Lewisia rediviva

FOOD: Bitterroot was a staple food for many tribes, but Europeans usually found it too bitter to enjoy. The starchy roots were collected just before the plants flowered, when the bitter, brownish-black outer layer could be easily peeled off by rubbing between the hands. Removal of the extremely bitter orange-red heart and storage for 1–2 years was said to reduce bitterness (though some people say that stored roots become increasingly bitter). Peeled, cored, washed roots were baked, steamed or boiled until soft and were eaten plain, but more often they were mixed with berries or powdered camas roots (p. 105), added to stews and used to thicken gravy. Extra roots were dried for a few days and stored for future use. A 50-lb (23-kg) bag of roots was considered enough to sustain a person through winter, and it took a woman 3–4 days to gather this amount. The brittle, white, dried roots were reconstituted by soaking and boiling, during which they would swell 5–6 times in volume and develop a jelly-like consistency.

MEDICINE: Various tribes used bitterroot tea to treat heart trouble and inflammation of the membrane that enfolds the lungs (pleurisy), to increase milk flow after childbirth and to purify the blood and relieve associated skin problems and diseases.

OTHER USES: The dried roots were an important trade item—a bag of bitterroot could be traded for a good horse. Bitterroot is Montana's state flower.

DESCRIPTION: Low, perennial herb with fleshy, club-shaped basal leaves ¹/₂–2" (1–5 cm) long from deep, fleshy taproots. Flowers deep to light rose-pink (sometimes whitish), with yellow to orange centers, 1⁵/₈–2¹/₂" (4–6 cm) across, with 12–18 lance-shaped petals and 6–9 pinkish, oval sepals, borne singly on ¹/₂–1¹/₄" (1–3 cm) tall stalks above a whorl of slender bracts, in April to July. Fruits oval capsules, with 6–20 dark, shiny seeds. Bitterroot grows in dry, open foothill, montane and subalpine sites from southern BC and Montana to New Mexico.

WARNING:
Dried bitterroot swells in the stomach, so it should not be eaten in large quantities. This beautiful wildflower is becoming increasingly rare as a result of overgrazing and human development. Unless it is extremely common in an area, it should be harvested only in an emergency.

Stonecrop & Roseroot

Sedum spp.

FOOD: Young leaves and shoots have been eaten raw or cooked, but older plants can become bitter. Sweetness also varies with the species, and roseroot is one of the most popular. A touch of garlic enhances the cucumber-like flavor. These plants have been eaten raw in salads and as a trail nibble, cooked as a hot vegetable, or added to soups and stews. The fleshy rootstocks have also been eaten, either boiled alone or with other vegetables, or pickled in seasoned vinegar. Siberian Inuit ate roseroot rootstocks with fat, after boiling them or fermenting them in water. These juicy plants can be a good source of liquid when water is not available.

MEDICINE: The plants are high in vitamins A and C. Because stonecrops are slightly astringent and mucilaginous, their juice or mashed leaves have been applied to wounds, ulcers, minor burns, insect bites and other skin irritations. Stonecrops have also been taken internally to treat lung problems and diarrhea.

Lance-leaved stonecrop (top); Roseroot (above)

Similarly, roseroot leaves were steeped in hot water or roots were boiled to make medicinal teas for relieving cold symptoms, for gargling to soothe sore throats and for washing irritated eyes.

DESCRIPTION: Hairless, succulent, perennial herbs with short, thick, alternate leaves on erect, 2–8" (5–20 cm) tall stems. Flowers small, 5-petaled, forming dense, flat-topped clusters, in May to August. Fruits compact clusters of 5 pointed capsules that split open along the upper inner edge.

Roseroot (*S. integrifolium* or *Tolmachevia integrifolia*) has strongly flattened, oval stem leaves and red to purple flowers and capsules. It grows on rocky subalpine and alpine slopes from Alaska to Colorado. Most *Sedum* species have yellow flowers and small, rounded stem leaves.

Lance-leaved stonecrop (*S. lanceolatum*) has cylindrical, finely bumpy, deciduous leaves and erect capsules. It grows on dry, rocky slopes in plains to alpine zones from the Yukon to New Mexico.

Lance-leaved stonecrop

119

Wintercress

Wild mustards
Brassicaceae

FOOD: The typical mustard bite of these plants varies with the species, habitat and season, but most have been used for food at some time. Tender young shoots provide a zesty addition to salads and sandwiches, and stronger plants are best used as flavoring in soups, salads and casseroles. Wild mustard plants were steamed in fire pits or boiled, sometimes with 2 changes of water to reduce the bitterness. The pods and seeds have a stronger taste than the plants and have been used as peppery additions to meats and salads. The seeds have traditionally been ground to powder and mixed with salt and vinegar to make sharp, peppery sauces (prepared mustards). The seeds of tumble-mustard were sometimes parched and ground into meal or flour, which was cooked into a mush, made into bread or used to thicken soups.

MEDICINE: These plants are rich in vitamins A, B and C and also contain considerable amounts of trace minerals. The seeds, which are said to stimulate the production of digestive juices, have been used for many years as an aid for digestion. Traditionally, mustard seed was used to relieve bronchial congestion by applying it to the chest as a mustard plaster (a paste of ground mustard and water sandwiched between 2 sheets of soft cloth). These plasters may help relieve congestion, but they should be used with caution (see Warning). Mustard seed has also been used in poultices for healing wounds, and it was sometimes taken in large quantities as a laxative (see Warning). Plants in the Mustard family contain isothiocyanate derivatives that may help to prevent cancer.

DESCRIPTION: Wide variety of wild mustards grow as weeds in the Rocky Mountains. Tall plants with pinnately divided leaves, yellow flowers and erect to ascending pods, containing a single row of seeds.

The **true mustards** (*Brassica* spp.) are recognized by their unusual, stout-beaked pods.

Tansy mustards (*Descurainia* spp.) have finely divided leaves covered with branched to star-shaped hairs, whereas most other mustards have simple hairs or are hairless.

Wintercresses (*Barbarea* spp.) have broadlobed leaves that clasp the stem, whereas **tumble-mustards** (*Sisymbrium* spp.) do not have clasping leaves. Of the white-flowered, weedy mustards, **peppergrasses** (*Lepidium* spp.) are easily recognized by their long, bottlebrush-like clusters of small, flattened, oval pods.

Most of these plants grow on open, disturbed ground in plains, foothills and montane zones throughout much of the Rockies.

WARNING:
Mustard plants are generally not poisonous, but their powdered seeds can be irritating, and mustard oil is one of the most toxic essential oils known. Ingestion of large quantities of mustard plants has caused poisoning of livestock. Mustard plasters redden the skin and prolonged contact (more than 10 or 15 minutes) raises blisters. Large quantities of mustard seed will irritate the digestive tract and cause vomiting. The thyroid is affected by isothiocyanate, which causes excessive secretions (hypertyroidism) and enlargement (goiter).

Wintercress (top); Peppergrass (above)

Bittercresses
Cardamine spp.

FOOD: All bittercresses are edible. They can be eaten raw, but some say they are better cooked in soups, stews and casseroles or served as a puree, like spinach. The delicate, slightly succulent plants add a refreshing, peppery flavor to sauces, salads and sandwiches. The pods of some species have been ground and mixed with vinegar to make a substitute for horseradish.

MEDICINE: Various bittercresses were used by native peoples as medicines. The pungent plants were used to treat fevers, colds, sore throats, headaches, heart palpitations, chest pains, gas, stomach upset and lack of appetite. Europeans used bittercress plants to prevent scurvy and to treat cramps, nervous hysteria, spasmodic asthma, urinary tract problems (especially gravel) and tumors. Dried flower clusters were believed to cure epilepsy in children. These plants are rich in vitamin C (and in the salts of iron and calcium), which would explain their effectiveness against scurvy, and, like many members of the Mustard family, they contain compounds that are believed to combat cancer.

OTHER USES: These plants are easily propagated by root division in autumn or from seed sown in spring. Some of the showier species are cultivated as ornamentals in Europe.

DESCRIPTION: Hairless herbs with basal and alternate leaves that are usually pinnately divided with a large leaflet at the tip. Flowers small, white, 4-petaled, forming elongating clusters, in April to July. Fruits slender, ascending pods $1/2–1^1/4$" (1–3 cm) long.

Pennsylvanian bittercress (*C. pensylvanica*) is an annual or biennial, with $1/8$" (3–4 mm) long petals and 5–11 linear-lance-shaped leaflets. It grows on moist foothill, montane and subalpine sites from the Yukon and NWT to Colorado.

Pennsylvanian bittercress (top)
Heart-leaved bittercress (above)

Two common perennial species have long, slender rootstocks.

Brewer's bittercress (*C. breweri*) has flowers with $1/8–1/4$" (3–7 mm) long petals and leaves with 3–7 broad leaflets. It grows on wet foothill and montane sites from BC to Wyoming.

Heart-leaved bittercress (*C. cordifolia*) has flowers with $1/4–1/2$" (7–12 mm) long petals and undivided, heart-shaped leaves. It grows in moist montane and subalpine sites from BC and Montana to New Mexico.

Watercress & Yellowcress

Rorippa spp.

Marsh yellowcress

FOOD: These plants have usually been eaten raw in salads and sandwiches, but they can also be steamed, boiled or stir-fried, like spinach. Common watercress is considered superior, but both are edible. The flavor is likened to peppery lettuce or radishes. Watercress fresh or dried is excellent in casseroles and stews and has also been used in gourmet soups and sauces. Sprigs of flowers can make a pretty peppery addition to salads. The seeds have been sprouted for use in salads or dried, ground and mixed with salt and vinegar to make water mustard.

MEDICINE: Carrot-and-watercress soup has been recommended for treating canker sores and blisters in the mouth. Watercress has also been used to treat a wide range of ailments, including boils, tumors, warts, baldness, eczema, scabies, fevers, flu, rheumatism, asthma, bronchitis, tuberculosis, goiter, nervousness and liver, heart and kidney problems. It was sometimes given to women during labor and was used as a contraceptive and as a cure for impotence. Watercress is rich in vitamins A, B_2, C, D and E and iodine. It was used for many years to prevent and cure scurvy. It also contains gluconasturtiin, which may help to protect smokers from lung cancer.

OTHER USES: The Romans used a mixture of watercress and vinegar to treat people with mental illnesses. From this use came the proverb, 'Eat cress and learn more wit.' The juice was said to clear the complexion of spots and pimples, and the plants were eaten to improve appetite.

WARNING:
Never eat raw plants from sites with polluted water. If you would not drink the water, either disinfect the watercress (e.g., soak it in water with water-purification tablets) or serve it cooked. Watercress should not be used medicinally by pregnant women, children or people with kidney inflammation, stomach ulcers or duodenal ulcers. Mustard-oil glycosides in watercress can cause intestinal and stomach upset if plants are eaten in large quantities or over long periods.

DESCRIPTION: Annual or biennial herbs with alternate, pinnately divided leaves with the largest leaflet at the tip. Flowers small, 4-petaled, forming elongating clusters, in March to October. Fruits sausage-shaped to rounded pods on spreading stalks.

Marsh yellowcress (*R. palustris* or *R. islandica*) is an erect, yellow-flowered, taprooted species, with oblong or oval, $1/8–3/8"$ (3–8 mm) long pods on equally long stalks. It grows on muddy sites in plains, foothills and montane zones from Alaska to New Mexico.

Common watercress (*R. nasturtium-aquaticum* or *Nasturtium officinale*) is a sprawling, white-flowered species that roots at its stem joints, producing bright green, tangled masses of shoots. Its relatively short-stalked pods are $1/2–1"$ (1–2.5 cm) long. This European plant has spread to quiet waters across temperate North America.

Shepherd's-purse

Capsella bursa-pastoris

FOOD: All parts of shepherd's-purse are edible, and these plants were once cultivated for greens. Young, crisp leaves have been used in salads, but usually plants were cooked in soups and stews. Shepherd's-purse is said to taste like a delicate cross between turnips and cabbages, and it has been recommended as a substitute for spinach. A pinch of baking soda in the cooking water helps to tenderize older plants. The slightly peppery pods and/or seeds can be used like mustard in cooked dishes, or they can be germinated to produce sprouts for salads and sandwiches. Some tribes crushed the dried pods, removed the chaff by winnowing, parched the seeds and ground them into a nutritious flour used in breads or mush. The roots, fresh and dried, have been used as a substitute for ginger. The fine, gray ash left by burned plants is high in sodium, potassium and other salts. It has been used as a salt substitute and tenderizer in cooked dishes.

MEDICINE: These plants are rich in vitamin C and have been reported to stimulate urination and stop bleeding. A strong decoction was used as drops for relieving earaches. Stems and leaves were steeped in hot water to make a tea for relieving headaches, stomach cramps, hemorrhoids and internal bleeding (especially bleeding of the uterus or kidneys). Shepherd's-purse was traditionally used during childbirth, to stop bleeding and aid delivery of the afterbirth by causing the muscles of the uterus to contract (an effect similar to that of oxytocin). The US National Cancer Institute reports that these plants may help to prevent cancer. In animal experiments, shepherd's-purse extracts have been shown to speed healing of stress-induced ulcers, fight chemically induced inflammation, increase urination, lower blood pressure, inhibit quivering contractions of the heart (ventricular fibrillation) and stimulate smooth muscles (especially in the intestines and uterus).

DESCRIPTION: Slender, annual herb with toothed or lobed basal leaves in rosettes and smaller, clasping stem leaves. Flowers white, $1/8$" (3–4 mm) across, 4-petaled, forming dense, rounded clusters, in March to July. Flower clusters elongate and bear flattened, triangular to heart-shaped pods $1/8$–$3/8$" (4–8 mm) long. Shepherd's-purse is a weed of disturbed, waste or cultivated ground in plains to subalpine zones from Alaska to New Mexico. It was introduced from Europe and has spread across North America.

WARNING:
The seeds have been known to cause blistering of the skin. Because these plants contain compounds that stimulate uterine contractions, they should not be eaten by pregnant women.

Field pennycress

Thlaspi arvense

ALSO CALLED: stinkweed.

FOOD: The young tender leaves have been eaten raw in salads, sandwiches and hors d'oeuvres, but they tend to have the characteristic bite of a mustard. Pennycress is too bitter and aromatic for most tastes, so it has usually been cooked in 1–2 changes of water and mixed with blander herbs or used for flavoring casseroles, soups and sauces. Mature plants are tough and strong-tasting. The seed pods have a peppery flavor and have been used as a mustard-like flavoring in soups and stews.

MEDICINE: Pennycress was historically used as an ingredient in the Mithridate confection (an elaborate preparation that was used as an antidote to poison), but by the early 1900s it was no longer used in medicine. The plants are high in vitamins C and G and also contain relatively large amounts of sulphur. Some reports have suggested that they may have healthful effects similar to those of sulphur and molasses. In some countries, the young plants were eaten to harmonize the internal organs and brighten the eyes. The seeds have been used to treat ophthalmia (inflammation of the eyeball) and lumbago (rheumatic pain in the lower back). Pennycress plants have a broad anti-microbial activity (against *Candida, Eschscherichia, Mycobacterium, Proteus, Pseudomonas, Staphylococcus* and *Streptococcus*), but their high mustard-oil content can be very irritating. In Eurasia, pennycress was considered an astringent, purifier, diuretic, stimulant and tonic, and it was also used to treat rheumatism.

WARNING:
Pennycress seeds contain the irritating substance, mustard oil (isothiocyanate). Cattle have been poisoned by eating hay with 25 percent or more pennycress. Even small amounts of field pennycress can taint the flavor of a cow's milk. People with sensitive skin may react to these plants when contact is accompanied by exposure to sunlight.

DESCRIPTION: Hairless, yellow-green annual herb with most leaves clasping the stem. Flowers white, 4-petaled, about 1/4" (6 mm) across, forming dense, rounded clusters, in April to August. These clusters greatly elongate and bear large (3/8–3/4" [9–18 mm] long), flat, round, heart-shaped pods that are broadly winged and notched at their tips. Field pennycress is a weed of disturbed, waste or cultivated ground in plains to subalpine zones from Alaska to New Mexico. It was one of the first weeds introduced to North America from Europe and has since spread across the continent.

Spider-flower

Cleome serrulata

ALSO CALLED: Rocky Mountain bee-plant.

FOOD: Spider-flower has an offensive, skunky smell and an alkaline taste, but both are said to disappear when plants are properly prepared. Tender young plants (up to a few inches tall) were boiled for a couple of hours or all day. Then the water was discarded and they were boiled in fresh water for an additional 2 hours. Sometimes meat was added to the mixture to make a stew, or the cooked plants were lightly salted and eaten with cornmeal porridge. Cooked plants were also drained and fried, or they were pressed dry (3 times), formed into cakes or small balls and dried for later use. The dried cakes/balls were soaked and then boiled in soups and stews or fried. The young flowers and seed pods were also eaten, and the seeds were ground to make flour. Spider-flower is said to have saved the Navajo from starvation several times. It was considered important enough to be named in songs with the 3 main cultivated plants: corn, pumpkin and cotton.

MEDICINE: The plants were dried, finely ground and mixed in water to make a medicinal tea for treating stomach problems. Also, fresh plants were wrapped in cloth and applied to the abdomen as a poultice.

OTHER USES: Tea, made by boiling a few seeds in water, was given to performers to strengthen their voices and give them 'good blood.' The cooled leaf tea was applied as a body deodorant, and the leaves were placed in shoes and moccasins to reduce odor. Spider-flower was sometimes used for sheep and horse feed. The young plants were used to make black paint for decorating pottery and other items. They were boiled to a thick, black mush and then dried in cakes that kept indefinitely. Later, the cakes were soaked in hot water until they reached the consistency of paint.

DESCRIPTION: Freely branched annual herb with stalked leaves divided into 3 oblong to lance-shaped, 5/8–2 3/4" (1.5–7 cm) long leaflets. Flowers pale pink to reddish-purple, about 3/4" (2 cm) across, forming showy, leafy, head-like clusters, in May to August. Clusters gradually elongate and produce 1 1/4–2 1/2" (3–6 cm) long, linear, cylindrical pods on 1/2–3/4" (1–2 cm) long stalks. Spider-flower grows on dry, open, often disturbed ground in plains, foothills and montane zones from southern Canada to New Mexico.

Silverweed

Argentina anserina

ALSO CALLED: *Potentilla anserina.*

FOOD: The flavor of these long, starchy rootstocks has been likened to that of parsnips, sweet potatoes or chestnuts. They were usually roasted, boiled or fried for a few minutes and then served as a hot vegetable or added to soups and stews. Silverweed roots were sometimes eaten raw, but they can be slightly bitter. They were said to taste best in the cold months, so they were usually collected in autumn or spring. Some roots were usually dried and stored for winter.

MEDICINE: Medicinal teas, made by steeping the leaves in hot water, were widely used to ease indigestion and menstrual cramps. Silverweed tea, sweetened with honey, was taken to relieve sore throats, and tea made with milk was used to treat diarrhea and dysentery. The leaves were also boiled to make mouthwash for soothing and healing sore gums and toothaches and to make ointment for treating skin problems. These plants are rich in tannin, which is very astringent—accounting for many of the medicinal uses of silverweed tea.

DESCRIPTION: Grayish-hairy, perennial herb with long, slender stolons. Leaves 4–8" (10–20 cm) long, pinnately divided into 11–25 leaflets interspersed with smaller leaflets. Flowers bright yellow, $5/8$–$3/4$" (1.5–2 cm) across, with 5 petals, 5 sepals and 5 smaller bractlets, borne singly, in May to August. Fruits corky-ridged, $1/16$" (1.5–2 mm) long seed-like achenes in dense clusters covered by the calyx. Silverweed grows on moist, open sites in plains, foothills and montane zones from the Yukon and NWT to New Mexico.

WARNING:
Silverweed can stimulate the uterine muscle, so pregnant women should not use this plant.

Wild strawberry (all images)

Strawberries

Fragaria spp.

FOOD: Each of these delicious, little berries seems to hold all the flavor of a large domestic strawberry. Wild strawberries are probably best enjoyed as a nibble along the trail, but they can also be collected for use in desserts and beverages. A handful of bruised berries or leaves, steeped in hot water, makes a delicious tea, served either hot or cold. Today, strawberries are preserved by freezing, canning or making jam, but traditionally they were sun-dried. The berries were mashed and spread over grass or mats to dry in cakes, which were later eaten dry or re-hydrated, either alone or mixed with other foods as a sweetener. Strawberry flowers, leaves and stems were sometimes mixed with roots in cooking pits as a flavoring. Wild strawberry is the original parent of 90 percent of cultivated strawberries.

MEDICINE: A vitamin C supplement can be made by covering fresh strawberry leaves with cold water, blending them to a pulp, and simmering the mixture for 15 minutes. Strain the following day and store in the refrigerator or freeze for later use. Strawberry-leaf tea is so high in vitamin C that the first-reported measurements were believed to be bogus. The leaf tea was used as a tonic and stomach cleanser, as a medicine for fevers, dysentery, diarrhea and kidney problems and as a wash for eczema, sores and other skin problems. It is said to be one of the best home remedies for diarrhea. Dried, ground leaves were used as a gentle disinfectant on open sores, applied as a powder or mixed with fat and used as a salve. Strawberries contain many quickly assimilated minerals (e.g., sodium, calcium, potassium, iron, sulfur and silicon), as well as citric and malic acids, and they were often taken to enrich the bloodstream. Strawberry-leaf tea, accompanied by fresh strawberries, was used as a remedy for gout, rheumatism, inflamed mucous membranes and liver, kidney and gall bladder problems. Stems and roots, collected late in the season, were boiled to make medicinal teas for healing sore throats and

> **WARNING:**
> Partially wilted leaves can contain toxins. Always use fresh or completely dried leaves. Some people develop a rash or hives after eating large quantities of strawberries.

mouth sores and for strengthening convalescents. The root tea was used to treat diarrhea and infant cholera. With the addition of yarrow root (p. 175), this tea was believed to cure insanity. Cooked strawberry plants were eaten to strengthen gums, fasten teeth, soothe inflamed eyes and relieve hayfever. A chemical, D-catechin, found in the leaves is said to inhibit histamine production. Although the leaf tea has little effect on its own, it seems to enhance the action of antihistamine drugs. Strawberries are a good source of ellagic acid, a chemical that is believed to prevent cancer.

OTHER USES: To remove tartar and whiten discolored teeth, strawberry juice was held in the mouth for a few minutes and then rinsed off with warm water. This treatment is most effective with a pinch of baking soda in the water. Large amounts of fruit in the diet can slow dental plaque formation. Strawberry juice, rubbed into the skin and later rinsed off with warm water, was used to soothe and heal sunburn.

DESCRIPTION: Low perennials with long, slender stolons. Leaves are 2–4" (5–10 cm) across, with 3 sharply toothed leaflets. Flowers are white, $5/8$–$3/4$" (1.5–2 cm) across, forming small, loose clusters, in May to August. Fruits are tiny seed-like achenes embedded in a red, fleshy receptacle (strawberry).

Wild strawberry (*F. virginiana*) has rather bluish-green leaflets, with the end tooth shorter than its adjacent teeth. It grows on well-drained sites in plains to subalpine zones from northern Canada to New Mexico.

Wood strawberry (*F. vesca*) has yellowish-green leaflets, with the end tooth projecting beyond its adjacent teeth. It grows on moist sites in foothills and montane zones from BC and Alberta to New Mexico.

Wood strawberry (top)
Wild strawberry (above)

Clovers
Trifolium spp.

FOOD: Clovers were occasionally used for food during periods of famine. All are high in protein and have been eaten raw or cooked. However, these plants are difficult to digest and can cause bloating in man and beast, so they should be used in moderation. Cooking helps to counteract this effect, and some native tribes dipped the leaves in salt water to reduce indigestion. The flowerheads, fresh or dried, have been used in teas, salads, soups and stews. In Ireland, the dry flowerheads and seed heads of white clover were ground into flour for making bread. Clover sprouts are said to be superior to alfalfa sprouts. The creeping stems and roots were highly valued by some tribes. These parts were cooked and served during feasts, with long, straight roots for dignitaries and short, gnarled ones for the common folk.

MEDICINE: In ancient Rome, clover seeds soaked in wine or clover plants boiled in water were used as antidotes for the poisonous bites and stings of snakes and scorpions. Clover tea was taken to treat coughs, fevers, sore throats, rheumatism and gout and was applied topically to treat skin diseases. Its high tannin content makes it rather astringent, and this astringency may explain some of these uses. Red clover has long been used as a blood purifier, reputed to stimulate the liver and to remove toxins from the blood. It has been shown to contain blood-thinning compounds related to

WARNING:
In autumn, red clover may appear normal but may contain toxic alkaloids. Some cattle have been poisoned by late-cut hay.

Alsike clover (top); Red clover (above)

coumarin, plus compounds (isoflavones) that are reported to be both estrogenic and cancer preventive. Tea made from red-clover flowers has been taken as a mild sedative and as a treatment for asthma, bronchitis and coughing. It has also been applied externally to treat athlete's foot, sores, burns and ulcers. Dried flowers were sometimes smoked to relieve asthma.

OTHER USES: Clovers provide excellent forage for cattle—hence the phrase 'living in clover' in reference to a luxurious lifestyle. These plants produce large amounts of high-quality nectar and are commonly planted as a bee crop. Clover is also a soil-enriching crop that adds nitrogen to the soil. Clover leaves were sometimes dried and smoked. Many people carry a 4-leaf clover (a leaf with 4 leaflets, occasionally caused by a mutation) for good-luck. Red clover is the state flower of Vermont.

Alsike clover

DESCRIPTION: Short-lived, perennial herbs with leaves divided into 3 oval leaflets. Flowers slender, pea-like, borne in dense, round heads throughout summer. Stems erect or spreading.

Red clover

Red clover (*T. pratense*) is the largest of the 3 common species, with plants 16–32" (40–80 cm) tall and leaflets ³/₄–2" (2–5 cm) long. It has large (¹/₂–³/₄" [12–20 mm] long), stalkless, usually red or deep pink flowers in dense, ³/₄–1¹/₄" (2–3 cm) wide heads with 1–2 short-stalked leaves at the base.

Alsike clover (*T. hybridum*) and **white clover** (*T. repens*) are smaller plants (usually less than 20" [50 cm] tall), with smaller (¹/₄–¹/₂" [7–11 mm] long), short-stalked, paler flowers in looser heads on leafless stalks. Alsike clover is relatively erect, with pinkish flowers and ³/₄–1⁵/₈" (2–4 cm) long leaflets, whereas white clover is a sprawling plant with rooting joints, mostly white flowers and ¹/₂–³/₄" (1–2 cm) long leaflets.

White clover

All 3 species were introduced from Europe and now grow wild on disturbed and cultivated ground across North America.

White clover

Alfalfa

Medicago sativa

FOOD: Tender young leaves have been added to salads and sandwiches, but usually alfalfa is used to make rather bland tea—described as, at best, resembling the taste of sun-dried grass clippings and, at worst, a coloring agent for hot water. Alfalfa is probably best used in a mixture with stronger-flavored herbs. Dried, powdered leaves have been sprinkled on cereal and other food as a nutritional supplement.

MEDICINE: Alfalfa is rich in vitamins A, B, C, E, K and P, calcium, potassium, phosphorus, iron and protein (about 18–19 percent). Some herbalists have recommended alfalfa tea for use by convalescents, pregnant women and people taking sulfa drugs or antibiotics. It has also been used for treating celiac disease (in combination with a proper diet), and it was said to stimulate appetite and weight gain, stimulate urination and stop bleeding. Many unproved claims have been made for alfalfa, including treatments for cancer, diabetes, alcoholism and arthritis. It contains the anti-oxidant tricin, as well as estrogenic isoflavones, which some scientists believe prevent cancer. Alfalfa has been shown to kill some fungi. In experiments with monkeys, it has reduced both blood cholesterol levels and fatty deposits (plaque) on artery walls, both of which are risk factors for stroke and heart disease.

OTHER USES: Alfalfa provides excellent forage for livestock, though ingestion of large amounts can cause bloating. It also produces abundant, high-quality nectar for bees. Alfalfa has been used as a commercial source of chlorophyll and carotene. To make a toothbrush, peel the root, cut it into 2.5" (10 cm) pieces, dry slowly and gently tap the end with a hammer to separate the fibers.

DESCRIPTION: Sweet-smelling, deep-rooted perennial with weak, reclining stems. Leaves divided into 3 oblong leaflets that are sharply toothed on the upper half. Flowers deep purple to bluish (rarely white), slender, pea-like, 1/4–3/8" (7–10 mm) long, borne in dense, 3/4–11/4" (2–3 cm) long heads, throughout summer. Fruits small, dark, net-veined pods tightly coiled in 1–3 spirals. Alfalfa was introduced from Eurasia as a forage plant and has spread to many roadsides and disturbed sites.

WARNING: Alfalfa plants contain saponin. When eaten in large quantities, this substance causes the breakdown of red blood cells in livestock and subsequent bloating. Alfalfa also interferes with vitamin E metabolism, and large amounts of alfalfa seed can cause a blood-clotting disorder in humans called 'pancytopenia.' Alfalfa sprouts contain the amino acid canavanine, which may cause the recurrence of lupus in patients in whom the disease has become dormant.

Common sweet clover

Melilotus officinalis

FOOD: The flowers and/or leaves can be used, fresh or dried, to make a pleasant, vanilla-flavored tea. Young leaves (gathered before flowering has begun) have been used raw in salads or boiled or steamed as a cooked vegetable. The seeds have been added to stews and soups as a flavoring, and crushed, dried leaves are said to give a vanilla-like flavor to pastries. These flowers and fruits are used to flavor Gruyère cheese in Switzerland.

MEDICINE: These plants are nutritious and rich in protein. Sweet-clover tea was traditionally used to treat a wide range of maladies, ranging from headaches, gas and nervous stomachs to painful urination, colic, diarrhea, painful menstruation, aching muscles and intestinal worms. The plants have also been used in poultices for treating swollen joints, inflammation, ulcers and wounds (especially in tender areas such as around the eye). Dried leaves of sweet clover were sometimes smoked to relieve asthma. Sweet clover contains quercetin, a substance used to treat fragile capillaries. These plants contain coumarin, which gives them their sweet, vanilla-like smell and flavor. Coumarin has been used to develop synthetic substances that prevent blood clotting in rats. Such anti-coagulants could be useful for treating blockage of the arteries to the heart muscle. People who were chilled from falling in water or being in heavy rain were given cold, sweet-clover tea, both as a drink and as a lotion.

OTHER USES: Sweet clover provides excellent forage for livestock and produces abundant, high-quality nectar for bees. It also increases the soil's nitrogen content.

DESCRIPTION: Sweet-smelling, $1\frac{1}{2}$–10' (50–300 cm) tall, annual or biennial herb with a strong taproot. Leaves divided into 3 oblong, $\frac{3}{4}$–$1\frac{1}{4}$" (2–3 cm) long leaflets edged with sharp teeth from near the base to the tip. Flowers yellow or white, pea-like, about $\frac{1}{4}$" (5 mm) long, forming branched clusters of long, slender spikes, in May to September. Fruits small ($\frac{1}{8}$" [3–4 mm] long), net-veined, mostly 1-seeded pods. Sweet clover was introduced from Eurasia and has spread to disturbed sites across North America. Until recently, the white variety was recognized as a separate species, *M. alba*.

WARNING:
If these plants are allowed to mold, their coumarin becomes dicoumarol, a chemical that can cause uncontrollable bleeding (an anti-coagulant).

Sweet-vetches

Hedysarum spp.

ALSO CALLED: licorice-root
• bear-root.

FOOD: Sweet-vetch roots were an important food for many tribes and were also enjoyed by trappers and settlers. Young roots have a sweet, licorice-like taste and were often eaten raw as a sweet treat. They were also boiled, baked or fried as a hot vegetable or added to soups and stews. Cooked sweet-vetch roots are said to taste rather like carrots. Some tribes enjoyed them dipped in grease. A hot drink was made by frying a small piece of root and then soaking it in hot water. When quantities permitted, roots were stored in buried caches, preserved in lard or oil, or dried for winter use. Sweet-vetch roots were usually collected in spring or autumn, because they become woody during summer.

OTHER USES: Pieces of root, softened at one end by chewing, were given to babies as pacifiers. Preserved roots were an important trade item among some tribes.

DESCRIPTION: Perennial, taprooted herbs with alternate leaves divided into 9–21 elliptic to oblong leaflets. Flowers pea-like, $1/2$–$3/4$" (1–2 cm) long, hanging in elongating clusters, in June to August. Fruits flattened pods constricted into 1–6 rounded segments.

Yellow sweet-vetch (top); Northern sweet-vetch (above)

Yellow sweet-vetch (*H. sulphurescens*) is an erect, leafy plant with prominently veined leaves (on the lower surface) and yellowish flowers. It grows on moderately dry to moist plains, foothill, montane and subalpine sites from Alberta and BC to Wyoming.

Most sweet-vetches have pink to reddish-purple flowers.

Alpine sweet-vetch (*H. alpinum*) has conspicuously veined leaflets, narrow-winged, 1/8–1/4" (3–6 mm) wide pods and small (mostly less than 5/8" [16 mm] long) flowers. It grows in moist, open montane, subalpine and alpine sites from Alaska to Wyoming.

Western sweet-vetch (*H. occidentale*) has prominently veined leaves, winged, 3/8–5/8" (9–14 mm) wide pods and larger (mostly greater than 3/4" [18 mm]) flowers. It grows in montane forests from Idaho and Montana to Colorado.

Northern sweet-vetch (*H. boreale*) has thick, obscurely veined leaflets. It grows in moist foothill, montane and subalpine sites from Alaska to New Mexico.

Northern sweet-vetch

WARNING:
There is some controversy over the edibility of northern sweet-vetch. In 1852, Sir John Richardson, the Arctic explorer, reported that all of his men who mistook northern sweet-vetch for the edible alpine sweet-vetch became ill. However, this story appears to be the only reference to the plant's toxicity. It is possible that toxicity varies with location, but this case of poisoning could also have resulted from confusion with another plant, such as a locoweed (p. 248). Sweet-vetch roots are best dug in spring, just as the shoots are coming up, which can make identification difficult. It is important to make sure that you have the right plant before you start eating, and northern sweet-vetch should be used with caution.

Yellow sweet-vetch

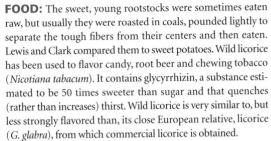

Wild licorice

Glycyrrhiza lepidota

FOOD: The sweet, young rootstocks were sometimes eaten raw, but usually they were roasted in coals, pounded lightly to separate the tough fibers from their centers and then eaten. Lewis and Clark compared them to sweet potatoes. Wild licorice has been used to flavor candy, root beer and chewing tobacco (*Nicotiana tabacum*). It contains glycyrrhizin, a substance estimated to be 50 times sweeter than sugar and that quenches (rather than increases) thirst. Wild licorice is very similar to, but less strongly flavored than, its close European relative, licorice (*G. glabra*), from which commercial licorice is obtained.

MEDICINE: Wild-licorice tea has been used for treating stomachaches, diarrhea, fevers (especially in children), stomach ulcers, asthma, arthritis and rheumatism. It was also taken to regulate menstrual flow and aid in the delivery of the afterbirth. The juice from chewing raw roots was held in the mouth to relieve toothaches, and leaves were steeped to make compresses or drops for curing earaches. Wild licorice roots are rich in mucilage, which makes them effective in soothing coughs and sore throats and healing ulcers. Some closely related species act as expectorants and also have proved to be as effective and longer lasting than codeine for suppressing coughs. These roots contain cortisone-like substances that are said to reduce inflammation without the dangerous side effects of steroids. Glycyrrhizin has been suggested as a safe sugar substitute for diabetics. European and Chinese species (*G. glabra* and *G. uralensis*) are among the most widely used medicinal plants in the world. Licorice has been shown to help heal and prevent stomach ulcers, to combat bacteria, viruses and yeasts, to reduce inflammation, and to lower levels of complexes that are formed in auto-immune diseases (systemic lupus erythmatosus).

OTHER USES: Chewed leaves were used as a poultice on horses with sore backs.

DESCRIPTION: Aromatic, glandular-dotted, perennial herb with leafy, 1–3^1/$_4$' (30–100 cm) tall stems from deep, spreading rootstocks. Leaves pinnately divided into 11–19 lance-shaped, ^{3}/$_4$–1^5/$_8$" (2–4 cm) long leaflets. Flowers yellowish-white to greenish-white, pea-like, about ^{1}/$_2$" (12 mm) long, forming dense, stalked clusters, in May to August. Fruits brown, bur-like pods, ^{3}/$_8$–^{5}/$_8$" (10–15 mm) long, covered with hooked bristles. Wild licorice grows in moist, well-drained sites, usually near water, in plains and foothills zones from BC and Alberta to New Mexico.

WARNING:
Large amounts of wild licorice can be toxic. Consumed in large amounts over time, it can act like a steroid hormone, raising blood pressure, increasing sodium retention and depleting potassium levels, which can cause water retention, elevated blood pressure, low energy levels, weakness and even death. Pregnant or nursing women, people with heart disease, high blood pressure, glaucoma or kidney or liver disease, and people taking hormones or digitalis should not use this plant.

Wild ginger

Asarum caudatum

FOOD: Wild ginger smells and tastes like commercial ginger (*Zingibar officinale*) and has been used in much the same way, but these 2 plants are not even distantly related. The leaves are more strongly flavored than the rootstocks, and they are generally milder than commercial ginger. The highly valued, widely used eastern species *Asarum canadense* has properties similar to those of its less-recognized western relatives. The rootstocks of wild ginger have been eaten fresh or dried as a ginger substitute, and the leaves have been used to make a fragrant tea. Rootstocks, boiled until tender and then simmered in syrup for 20–30 minutes, are said to make excellent candy. They have also been pickled in brandy or dried, ground and used as a substitute for commercial ginger.

MEDICINE: Candied rootstocks were used to relieve coughing and stomach problems. The leaves have antifungal and anti-bacterial properties, and they were used as poultices on cuts and sprains. Rootstocks were boiled to make medicinal teas for treating indigestion and colic. A stronger decoction was taken by women as a contraceptive and was used as drops for curing earaches. The leaf tea is said to stimulate sweating and increase secretions from the tear ducts, sinuses, mouth, stomach lining and uterus. It has been taken to relieve fevers, gas, stomach upset and slow, crampy menstrual periods, and to cleanse the skin when treating measles, chicken pox, rashes and acne. Wild ginger contains aristolochic acid, an anti-tumor compound.

OTHER USES: Wild ginger in the bedding of ill or restless babies was believed to have a quieting, healing effect. Dried, powdered leaves were used as a deodorant. Some wild ginger plants contain a potential slug-repellent. Plants growing in areas with slugs produce chemicals that kill or repel these pests, whereas plants in slug-free habitats do not.

WARNING:
Wild ginger should not be used by pregnant women. Large doses can cause nausea. People with sensitive skin may develop rashes from handling the fuzzy leaves. This unusual little plant is becoming less and less common, as its habitat disappears. Please leave it in its natural environment. Commercial ginger is readily available to use in its place.

DESCRIPTION: Trailing, often matted herb with 2 shiny, dark green, heart-shaped to kidney-shaped leaves 2–4" (5–10 cm) wide, smelling strongly of lemon-ginger when crushed. Flowers purplish-brown to greenish-yellow, bell-shaped, with 3 slender-tailed, petal-like sepals, single, at ground level (often hidden beneath the leaves), in April to July. Fruits inconspicuous, fleshy capsules. Wild ginger grows in moist, shaded foothill and montane sites from BC to Idaho and Montana.

Round-leaved yellow violet

Violets

Viola spp.

FOOD: All violets, including garden varieties such as Johnny-jump-ups and pansies, are edible. Most leaves are tender and sweet and make an excellent salad green or trail nibble, but they can also be cooked as potherbs or thickeners. The flowers provide pretty, edible garnishes (fresh or candied) for salads and desserts and delicate flavoring and/or coloring for vinegar, jelly, syrup, jams and preserves. The leaves and/or flowers have also been steeped in hot water to make tea or fermented to make wine.

MEDICINE: Violet plants are rich in vitamin A (richer than spinach) and vitamin C ($1/2$ cup can be equivalent to 4 oranges), and some contain as much as 4000 ppm salicylic acid (similar to aspirin). The flowers have significant amounts of rutin, a compound that strengthens capillary blood vessels. Violet plants have a laxative effect, which is said to be stronger in yellow-flowered species. They have been used to make medicinal teas for treating bronchitis, asthma, heart palpitations and fevers; gargles and syrups for relieving sore throats and coughs; and poultices, salves or lotions for treating bruises, rashes, boils and eczema. Violets are reported to have a mild hormone-regulating action, and early blue violet leaves were used by some women to ease labor. The ancient claim that violet leaves will cure cancer has not been substantiated. However, animal studies have shown that violet plants stimulate urination and that they may be useful in treating rashes. Violet roots were sometimes used to induce vomiting in poison victims.

OTHER USES: Violet flower tea was used as a substitute for litmus paper. Mashed violet leaves were burned like incense to ward off disease. Violet flowers are a popular fragrance for perfumes and have been used in many cosmetics. Illinois, New Brunswick, New Jersey, Rhode Island and Wisconsin all have wild violet as their state/provincial flower.

Early blue violet

WARNING:
Use violets in moderation. Some leaves contain saponin, which cause digestive upset in large quantities. Only the leaves, stems and flowers should be eaten. Violet roots, rootstocks, fruits and seeds contain toxins that can cause severe stomach and intestinal upset, as well as nervousness and respiratory and circulatory depression.

DESCRIPTION: Low, perennial herbs (mostly less than 8" [20 cm] tall) with basal or alternate, usually heart-shaped leaves. Flowers with 3 spreading lower petals, 2 backward-bending upper petals and a basal spur, single, in May to July. Fruits small capsules that shoot seeds out explosively.

Early blue violet (*V. adunca*) has white-throated, blue to deep violet flowers borne in the axils of stem leaves, and each leaf stalk has 2 slender, toothed lobes (stipules) at its base. It grows in foothill, montane and subalpine sites from northern Canada to Colorado.

Canada violet (*V. canadensis*) is a larger (4–16" [10–40 cm] tall) species with broad, heart-shaped stem leaves and purple-lined, pale violet to white flowers. It grows in foothill and montane woodlands from BC and Alberta to New Mexico.

Round-leaved yellow violet (*V. orbiculata*) has low-lying, basal leaves and yellow flowers marked with purplish lines. It grows in moist foothill, montane and subalpine woods from BC and Alberta to Idaho and Montana.

Canada violet

Yellow montane violet or **yellow prairie violet** (*V. nuttallii*) has tapered (not heart-shaped) leaves and grows in dry plains, foothill and montane sites from BC and Alberta to New Mexico.

Early blue violet

Scarlet globemallow

Sphaeralcea coccinea

FOOD: The roots were sometimes chewed to reduce hunger when food was scarce.

MEDICINE: Scarlet globemallow roots were chewed and then laid on sores and wounds to aid healing and stop bleeding. The leaves are slimy and mucilaginous when crushed, and they were chewed or mashed and used as poultices or plasters on inflamed skin, sores, wounds and aching or blistered feet. Leaves were also used in lotions to relieve skin diseases. Dried, ground leaves were dusted on sores or were mixed with water to make a paste for use as a protective, drawing poultice. Fresh leaves and flowers were chewed or made into teas for relieving hoarse or sore throats and upset stomachs. Whole plants were used to make a sweet-tasting tea that made distasteful medicines more palatable. This tea was also said to reduce swellings, improve appetite, relieve upset stomachs and strengthen voices.

OTHER USES: Tea made from scarlet globemallow was used as a remedy for diseases caused by witchcraft. In some tribes, the plants were chewed to a paste and rubbed onto skin for protection from scalding. The person was then able to amaze observers by dipping his hand into boiling water to take out objects such as hot meat, without being burned. The fuzzy leaves have been recommended as a soothing shoe-liner. These attractive plants can be propagated from seeds or cuttings or by dividing their roots.

DESCRIPTION: Low, spreading, perennial herb, grayish, with dense, star-shaped hairs, often forming patches from spreading rootstocks. Leaves alternate, 3/4–2" (2–5 cm) long, palmately cut into 3–5 wedge-shaped segments arranged like fingers on a hand (palmate). Flowers orange to brick-red, about 3/4" (2 cm) across, with 5 broad, shallowly notched petals and 5 reddish, hairy sepals, forming small clusters, in May to August. Fruits small (1/8" [3 mm]), 1-seeded, wedge-shaped segments arranged like wedges on a fuzzy wheel of cheese. Scarlet globemallow grows in dry, often disturbed, open plains, foothill and montane sites from southern BC and Alberta to New Mexico.

St. John's-worts

Hypericum spp.

MEDICINE: Although common St. John's-wort has been the most widely used, western St. John's-wort contains similar compounds, but in lower concentrations. St. John's-wort is rich in tannin, and it was used medicinally by native peoples and Europeans for treating many ailments, including diarrhea, tuberculosis, bladder problems and worms. The fresh flowers were steeped in water, alcohol or oil to make washes and lotions for treating wounds, sores, cuts, bruises and ulcers. The tea was taken to treat bladder problems, dysentery, diarrhea, worms and depression. Experiments have shown St. John's-wort extracts to have sedative, anti-inflammatory and anti-bacterial effects (effective against *Mycobacterium tuberculosis* and *Staphylococcus aureus*). The glandular dots contain hypericin, which has been used as a sedative and an anti-depressant for treating sadness, frustration, anxiety, irritability, insomnia and general blues and 'grouchies.' Hypericin inhibits the enzyme monoamine oxidase, thereby increasing the activity of nerve-impulse transmitters in the brain that are critical to mood and emotional stability. Oil extracts have been given internally to treat chronic stomach inflammation and stomach ulcers. They have also been applied externally to heal cuts, scrapes and mild burns and to relieve sciatica, back spasms, neck cramps and associated

St. John's-worts: Common (top); Western (below)

stress headaches. Two anti-viral agents in these plants (hypericin and pseudohypericin) are being studied for use in the treatment of AIDS.

WARNING:
Either taken internally or applied externally, hypericin is reported to cause light-sensitive rashes and hives on rare occasions. White or unpigmented cattle, sheep, horses and rabbits have developed severe light-sensitive skin reactions and died from eating large quantities of these plants. In many areas, common St. John's-wort is a troublesome weed.

DESCRIPTION: Hairless, perennial herbs with opposite, stalkless leaves and bright yellow flowers, both usually dotted along their edges with tiny, black glands. Flowers about ³/₄" (2 cm) across, 5-petaled, with 75–100 stamens gathered into 3 bundles, borne in leafy, open clusters, in June to September. Fruits small capsules.

Common St. John's-wort or **klamath weed** (*H. perforatum*) has slender sepals (3–5 times longer than wide) and lance-shaped leaves. This European weed now grows on disturbed ground at lower elevations from southern BC to Colorado.

Western St. John's-wort (*H. scouleri*) has broader sepals (less than 3 times longer than wide) and oval leaves. It grows on moist, open foothill, montane, subalpine and alpine slopes from BC and Alberta to New Mexico.

141

Prickly-pear cacti
Opuntia spp.

FOOD: Prickly-pear cacti were widely used for food, though the fruits of these species are smaller and less fleshy than those of their southern relatives. The flavor ranges from bland to sweet to sour and has been likened (at best) to sweet pomegranates. Spines were peeled off, burned off, picked off (with fingers protected by deerskin tips), or removed by sweeping piles of fruit with sagebrush (p. 87) branches. The fruits were then split to remove the seeds and eaten raw (alone or with other fruits), cooked in stews and soups as a thickener, or dried for later use. More recently, the sweet flesh has been added to fruit cakes or canned in fruit juice. Berries can also be boiled whole and strained to make jellies or syrups. Dried seeds were added to soups and stews or were ground into flour and used as a thickener. Raw cactus stems have been likened to cucumber, but they were usually eaten only when there was a shortage of food. Young segments were boiled and peeled to remove their spines, and the pulpy flesh was fried. Alternately, roasted or pit-cooked stems were simply squeezed until the edible inner part popped out. Cactus stems have also been pickled or candied.

MEDICINE: The peeled mucilaginous stems were used for dressing wounds or were mashed and placed on aching backs. Stems were also boiled to make a medicinal drink for relieving diarrhea and lung problems and for treating people who could not urinate. More recently, studies have suggested that the juice may be effective in lowering blood sugar levels in diabetics, especially those with chronic hyperglycemia.

Brittle prickly-pear cactus (top)
Plains prickly-pear cactus (above)

WARNING:
Always protect your hands with gloves when collecting these spiny plants.

OTHER USES: When forage was limited, the spines were singed off and cactus stems were fed to livestock. Split stems were placed in containers of muddy water, where they exuded large amounts of mucilage, which cleared the water and made it drinkable. Freshly peeled stems were sometimes rubbed over painted hides to fix the colors.

DESCRIPTION: Spiny, perennial herbs, 2–12" (5–30 cm) tall, with thick, fleshy, segmented stems. Leaves reduced to spirally arranged, star-burst clusters of short bristles and rigid, barbed, $1/2$–2" (1–5 cm) long spines. Flowers yellow, broadly bell-shaped, with many thin, overlapping petals, solitary at branch tips, in May to June. Fruits fleshy (though often rather dry), seedy, somewhat spiny, oval berries $5/8$–1" (1.5–2.5 cm) long.

Plains prickly-pear cactus (*O. polyacantha*) has slightly barbed spines in brown-wooly or hairless tufts on large (often 2–4 $3/4$" [5–12 cm] long), flattened stem segments that do not break apart easily.

Brittle prickly-pear cactus (*O. fragilis*) has strongly barbed spines in white-wooly tufts on smaller (less than 2" [5 cm] long), rounded stem segments that detach easily.

Both species grow on dry, open ground in plains to foothills zones from southern BC and Alberta to New Mexico.

Plains prickly-pear cactus (both photos)

Brittle prickly-pear cactus

Perennial eveningstars

Mentzelia spp.

FOOD: The oily seeds were gathered by some tribes, dried, parched and ground into a nutritious meal for making pinole or mush and for adding to soups and stews. The roots were sometimes chewed to relieve thirst.

MEDICINE: Smooth-stemmed eveningstar was one of the oldest and most respected medicines of the Cheyenne. Its roots were gathered early in the year, before the plants had flowered. Because they were considered such a powerful medicine, they were always mixed with other herbs and were never used alone. Smooth-stemmed eveningstar was considered especially good for relieving earaches and pain from rheumatism and arthritis. It was also used to make medicinal teas and salves for treating serious contagious diseases such as mumps, measles and smallpox.

OTHER USES: These giant, luminous flowers make a wonderful addition to a wildflower garden, especially in dry regions with poor soil, but their beauty remains hidden most of the time. Eveningstar flowers close tightly during the day and open at dusk to attract night-flying pollinators.

DESCRIPTION: Rough, grayish, perennial (sometimes biennial) herbs covered in barbed hairs, with fleshy, brittle leaves that readily catch onto passers-by and break off. Flowers large (about 4" [10 cm] wide), fragrant, with 5 or 10 spreading, pointed petals around a star-burst cluster of stamens, borne singly or in small clusters, in July to September. Fruits oblong capsules, $5/8$–$1 5/8$" (1.5–4 cm) long.

Giant eveningstar (*M. decapetala*) has creamy to yellowish flowers with 10 'petals' (5 true petals and 5 petal-like stamens), and its seeds lack wings. It grows on dry, open slopes and roadsides in plains, foothills and montane zones from southern Alberta to New Mexico.

Smooth-stemmed eveningstar (*M. laevicaulis*) has 5-petaled, lemon-yellow flowers and winged seeds. It grows on dry, open sites in plains, foothills and lower montane zones from southern BC and Montana to Wyoming.

Giant eveningstar (top)
Smooth-stemmed eveningstar (above)

Dogbanes

Apocynum spp.

MEDICINE: Dogbane roots were mashed and applied as poultices or were used to make teas and tinctures that were applied as lotions. These preparations were said to induce sweating, act as a counter-irritant, and stimulate blood flow to the skin. They were sometimes applied to the scalp to increase hair growth. Berry and root extracts were taken internally to relieve constipation, to induce sweating and vomiting, to stimulate the heart, to relieve insomnia, and to expel intestinal worms. However, this use could be very dangerous (see Warning). Dogbanes were used to treat headaches, indigestion, rheumatism, liver disease and syphilis. These plants have a more powerful action on the heart than digitalis, and they are also irritable to the digestive tract. Some of their glycosides are said to have anti-tumor properties.

OTHER USES: Mature stems were soaked in water to remove the tough outer fibers, which were then rolled against the leg to make thread. This thread was said to be stronger and finer than cotton thread. Three strands plaited together were used to make bowstrings, and the cord was also made into fishing nets and net bags. Dogbane cord was used to make rabbit nets, up to 1 mi (1.5 km) long, which were used in communal rabbit hunts. These nets were often inherited, and the owner was the recognized leader of the hunt. Indian hemp was preferred as a source of fiber because of its longer, tougher fibers. The milky sap (latex) contains rubber.

Spreading dogbane (top)
Indian hemp (above)

DESCRIPTION: Erect, perennial herbs with milky sap and opposite, sharp-pointed, short-stalked leaves. Flowers bell-shaped, 5-petaled, sweet-smelling, forming showy, branched clusters, in June to September. Fruits hanging pairs of slender pods (follicles), splitting down 1 side to release many small, silky-parachuted seeds.

WARNING:
Dogbanes contain glycosides that stimulate the heart and increase blood pressure. Ingestion has resulted in sickness and death. These compounds can also raise blisters on sensitive skin, but most people handle dogbane plants with impunity.

Spreading dogbane (*A. androsaemifolium*) is 8–28" (20–70 cm) tall, with oval to oblong leaves and $1/4$" (5–7 mm) long, pink flowers. It grows on well-drained plains to subalpine sites from Alaska to New Mexico.

Indian hemp (*A. cannabinum*) is 1–3$1/2$' (30–100 cm) tall, with narrower leaves and $1/8$" (2–4.5 mm) long, greenish-white to white flowers. It grows in moist plains and foothill sites from BC and Alberta to New Mexico.

145

Showy milkweed

Asclepias speciosa

FOOD: Young shoots, unopened flower buds and immature seed pods were eaten raw (see Warning), boiled, or fried in oil (with or without batter). Milkweed was often added to soups and stews to thicken broth and tenderize meat. It was usually necessary to first cover the plants with boiling water, return the water to a boil, discard the water and repeat this process 2–3 times to reduce bitterness. Young plants (4–6" [10–15 cm] tall) could be used like asparagus and young firm pods ($^{3}/_{4}$–1$^{1}/_{4}$" [2–3 cm] long) like okra. The sweet buds and flowers were boiled down to make a thick syrup or brown sugar. They were also made into preserves. Milkweed seeds were occasionally eaten.

MEDICINE: The milky sap was traditionally used to treat skin problems. It was applied to cuts and burns and to a wide range of infections and irritations, including warts, moles, ringworm, poison-ivy rash (p. 237), measles, corns and calluses. The plant tips were boiled to make a wash for treating blindness. The powdered roots were boiled to make medicinal teas for use as a sedative and as a treatment for stomachaches and asthma. Fresh roots were boiled to make teas for treating bowel problems, kidney problems, water retention, rheumatism, intestinal worms, asthma and venereal disease and for use as a temporary contraceptive. Seeds were boiled to make a solution for drawing poison from snake bites, or they were powdered and added to salves for treating sores.

OTHER USES: Dried juice from broken stems provided chewing gum. The milky juice leaves a stain that can persist for days or weeks, and it was sometimes used to temporarily brand livestock. The seed silk was used to stuff pillows, mattresses and comforters and was even woven (with other fibers) into cloth.

DESCRIPTION: Robust, grayish-hairy herb with milky sap (latex) and hollow, hairy stems. Leaves opposite, 4–10" (10–25 cm) long, with a pinkish midrib and conspicuous side veins. Flowers pink or whitish to greenish-purple, about 1" (2.5 cm) long, with 5 backward-bent petals below 5 erect, horn-like appendages, forming rounded clusters, in May to July. Fruits single or paired, soft-spiny pods (follicles) 2$^{3}/_{4}$–4" (7–10 cm) long, containing flat seeds with parachutes of silky hairs. Showy milkweed grows on open, loamy ground in plains, foothills and montane zones from southern BC and Alberta to New Mexico.

WARNING:
The milky sap contains poisonous cardiac glycosides. Livestock have been poisoned, but usually animals avoid these plants. The toxins are destroyed by heat, so milkweeds should always be cooked before they are eaten. The roots are considered poisonous. Narrow-leaved milkweeds are more toxic than broad-leaved species.

Monument plant

Frasera speciosa

ALSO CALLED: green gentian
• *Swertia radiata.*

FOOD: The fleshy roots have been used as a vegetable, raw, roasted or boiled. They are said to be best mixed with other roots and herbs, either in salads or in soups and stews.

MEDICINE: Hunters mixed the dried leaves with mountain tobacco (*Nicotiana tabacum*) and smoked the mixture in corn husk cigarettes or hunting pipes to gain strength and clear the mind. Lost hunters would smoke their pipes believing that they could then think clearly and find their way back to camp. Like its relatives, the gentians, monument plant has been used as a digestive stimulant, but it is said to be more energetic and irritating in its effects. The roots were used to make bitter tonics for stimulating secretions and contractions of the stomach and small intestine. The root powder has also been used as a fungicide for treating athlete's foot, jock itch, etc. Mixed with lard, it has been used to make salves for killing lice and scabies. Alcohol extracts of the roots have been used to treat ringworm, but these extracts can irritate the skin, especially of children.

OTHER USES: The Navajo considered monument plant a male plant and great mullein (p. 157) its corresponding female plant. The leaves of both plants were ground together and used to make a tea that was rubbed on the bodies of hunters and their horses to give them strength during long expeditions.

WARNING:
The eastern species, *F. caroliniensis* has been used to stimulate vomiting and diarrhea. Monument plant could have this effect if large quantities are eaten, but it is said to be much milder and safe for use in moderation.

DESCRIPTION: Large, pale green, biennial herb with single, 1–6¹/₂' (0.3–2 m) tall stems. Leaves 10–20" (25–50 cm) long, basal in the first year, in whorls of 3–5 in the second year. Flowers yellowish-green with 4 large, purplish markings and fringes of long hairs near the center, about 1¹/₄" (3 cm) across, 4-petaled, forming elongated clusters, in June to August. Fruits slightly flattened capsules ³/₄–1" (2–2.5 cm) long. Monument plant grows on moderately dry slopes in plains to sub-alpine zones from Montana to New Mexico.

Mountain bog gentian

Gentians

Gentiana spp. & Gentianella spp.

FOOD: Before the widespread introduction of hops, gentians were occasionally used in Europe, with other bitter herbs, for brewing beers.

MEDICINE: Gentian roots (and occasionally the flowers) have been used in many forms (powders, teas, syrups and alcohol extracts) to treat many ailments (fever, indigestion, jaundice, fluid accumulation, skin diseases and gout), but usually they have been used in bitter tonics to aid digestion. 'It [gentian] is one of the best strengtheners of the human system and is an excellent tonic to combine with a purgative to prevent its debilitating effects. It is of extreme value in jaundice and is prescribed extensively. Gentian possesses febrifuge, emmenagogue, anthelminthic and antiseptic properties and is also useful in hysteria, "female weakness," etc. A tincture made of 2 oz (60 ml) root, 1 oz (30 ml) dried orange peel and 1/2 oz (15 ml) cardamom seeds in a quart of brandy is an excellent stomach tonic and is efficacious in restoring appetite and promoting digestion' (Grieve 1931). Present-day herbalists still recommend gentian-root tea as one of the best vegetable bitters for stimulating appetite, aiding digestion, relieving bloating and flatulence and preventing heartburn. The bitter principles are also said to normalize the functioning of the thyroid gland.

OTHER USES: Many gentians are both beautiful and hardy, and they make excellent subjects for wildflower gardens.

DESCRIPTION: Hairless herbs with opposite, essentially stalkless leaves. Flowers usually blue to purple, tubular to bell-shaped, tipped with 4–5 erect to horizontal lobes, borne in June to September. Fruits small capsules.

WARNING:
Overdoses cause nausea and vomiting, but only the most grimly resolute could manage to eat large amounts of these extremely bitter plants. Pregnant women and people suffering from stomach or duodenal ulcers or very high blood pressure should not use these plants.

Northern gentian or **felwort** (*Gentianella amarella*) is an annual or biennial herb with branched stems. Its purplish-blue to pale yellow-and-blue flowers have fringed throats and form clusters of large and small, ³/₈–⁵/₈" (10–15 mm) long flowers in June to September. Northern gentian grows in moist foothill to alpine sites from the Yukon and NWT to New Mexico.

Rocky Mountain fringed gentian (*Gentianopsis detonsa*) has ³/₄–2" (2–5 cm) long, purplish-blue flowers with 4 broad, widely spreading, fringed lobes, and its calyx lobes have glossy, purple central ridges. It grows on moist montane to alpine slopes from Idaho to New Mexico.

Mountain bog gentian (*Gentiana calycosa*) has solitary, 1¹/₄–1⁵/₈" (3–4 cm) long, usually deep purplish-blue flowers with 5 blunt, spreading lobes and 5 pleated, fringed clefts. It grows in moist, open montane to alpine sites from BC and Alberta to Montana.

Northern gentian

Rocky Mountain fringed gentian

Western blue flax

Linum lewisii

ALSO CALLED: wild blue flax
• *Linum perenne.*

FOOD: Wild flax seeds are very rich in oil. They were gathered by some tribes, dried, roasted and ground into flour or meal. The seeds were also cooked with other foods as an oil-rich supplement. Flax seeds should not be eaten raw (see Warning). The annual European species, common flax (*L. usitatissimum*) is cultivated for its seeds. The seeds are sold commercially and are becoming increasingly popular as a healthy source of oil (see Medicine).

MEDICINE: Oil from flax seeds has been used for centuries in Europe for treating skin problems. The crushed seeds were applied to rashes and boils. Flax seeds contain 'healthy' oils, with both alpha-linoleic and *cis*-linoleic essential fatty acids. These have been shown to lower blood fat (lipid) and cholesterol levels, increase the local immune function of prostaglandin, and reduce clotting. Flax seed oil has been used to treat a wide range of ailments, ranging from breast and colon cancer to lupus-related kidney disease, psoriasis and eczema. The seeds are a good source of dietary fiber and also produce a sticky, glue-like substance (mucilage) when soaked in water for 20 minutes or more. Crushed flax seed is taken internally as a mild laxative with soothing effects on mucous membranes. Flax seed has also been applied externally to soothe and heal burns and boils. The leaves were boiled to make a medicinal tea for treating heartburn.

OTHER USES: The stems contain long, tough fibers (similar to those found in common flax) that were used in thread, cords and strings for making fishing lines and nets, snowshoes and some parts of mats and baskets. Common flax is cultivated for its fiber, which is used to make linen. Flowering plants were sometimes boiled to make a wash for the face and head, because the wash was believed to produce beautiful skin and hair. Western blue flax is used as an ornamental perennial, producing flowers for most of the summer and early autumn.

DESCRIPTION: Slender, gray-green, perennial herb with many linear, $1/2$–$1^1/4$" (1–3 cm) long leaves. Flowers pale blue, about $3/4$–$1^1/4$" (2–3 cm) across, with 5 fragile petals, soon fading, opening singly in few-flowered clusters, in May to August. Fruits round capsules, on curved stalks. Western blue flax grows on dry plains, foothill and montane slopes from Alaska to New Mexico.

WARNING:
Eating immature seed pods or large amounts of mature seeds can cause toxic reactions. Although livestock have been poisoned by eating flax, no cases of human poisoning have been reported.

Skullcaps

Scutellaria spp.

MEDICINE: Skullcaps contain a flavonoid called 'scutellarin' that has sedative and antispasmodic properties. Warm tea made from the dried leaves of this plant has been used by herbalists for over 250 years to treat hysteria, neuralgia, epilepsy, multiple sclerosis, Saint Vitus' Dance and convulsions. Some native peoples used it as a sedative and as a medicine to promote menstruation. In the 1700–1800s, skullcap was famed for its ability to cure the bites of mad dogs (rabies), but this ability was later disclaimed. More recently, it has been used to relieve nervous headaches, sciatica, shingles and the tension and irritability of premenstrual syndrome, to prevent epileptic seizures, and to help wean addicts from barbiturates, Valium and meprobamate. Similarly, skullcap has been mixed with American ginseng (*Panax quinquefolius*) to treat people suffering from the delirium tremens of alcoholism.

DESCRIPTION: Perennial herbs with weak, 4-sided stems and opposite, ³/₄–3" (2–8 cm) long, blunt-toothed leaves. Flowers blue to bluish-purple, trumpet-shaped, with a hooded upper lip, a broad lower lip and a 2-lipped calyx with a small bump on the upper side. Fruits 4 tiny, bumpy nutlets, produced in July to October.

Marsh skullcap (*S. galericulata*) has relatively few, large (⁵/₈–³/₄" [15–20 mm] long) flowers in its upper leaf axils (1 per leaf). It grows in moist plains, foothill and montane sites from the southern Yukon and NWT to New Mexico.

Blue skullcap (*S. lateriflora*) has many, smaller (¹/₄" [6–8 mm] long) flowers in long clusters from the leaf axils. It grows in moist areas in the plains and foothills zones from southern BC to New Mexico.

> **WARNING:**
> Although it is not considered poisonous, skullcap should be used in moderation. Too much can cause excitability and wakefulness, with giddiness, stupor, confusion and twitching.

Marsh skullcap (top 2 photos)
Blue skullcap (above)

Swamp hedge-nettle

Stachys palustris

ALSO CALLED: woundwort.

FOOD: The plump, crisp rootstocks, collected in autumn, are said to have a rather agreeable flavor. They have been eaten raw, boiled, baked or pickled. They were also dried, ground and used to make breads in times of food shortage. Young shoots can be cooked as a vegetable (like asparagus), but they are said to have a disagreeable smell. The flowers are also edible and some tribes ate the seeds.

MEDICINE: The leaves were bruised or soaked in water and applied as poultices to stop the bleeding of wounds and open sores. The leaves and roots have also been used in poultices to reduce pain and inflammation associated with sprains, swollen joints and headaches. Fresh roots were chewed alone or were soaked in alcohol that was then gargled to relieve sore throats. Medicinal teas made from these plants have been used to treat inflammation of the bladder and urethra, migraine headaches, headaches from eye strain and hangovers, sprains and inflamed joints. However, there is no scientific evidence to support these uses.

OTHER USES: Swamp hedge-nettle produces a yellow dye.

DESCRIPTION: Glandular-hairy, perennial herb with erect, 4-sided stems bearing opposite, 1⅝–3" (4–8 cm) long, blunt-toothed leaves. Flowers purplish-pink to whitish, mottled with darker spots, widely funnel-shaped, with a broad, concave upper lip and a flared, 3-lobed lower lip, ⅜–⅝" (10–15 mm) long, borne in spike-like clusters of few-flowered whorls, in July to August. Fruits dark brown nutlets, about 1/16" (2 mm) long, in clusters of 4. Swamp hedge-nettle grows on moist plains and foothill sites from the southern Yukon and NWT to New Mexico.

WARNING:
These plants are called 'hedge-nettles' because of their stiff, bristly hairs. However, unlike true nettles, they do not have stinging hairs, so they are relatively safe to touch.

Giant-hyssops

Agastache spp.

FOOD: The seeds of both species can be eaten raw or cooked. The leaves have been used to flavor soups, stews and other hot dishes. They can also be steeped in hot water to make a delicate anise-flavored tea, best when brewed weakly. The quality of the flavor of these plants can vary greatly from one population to the next.

MEDICINE: Giant-hyssop leaves have been used to make pleasant medicinal teas for treating coughs, colds and fevers and for relieving chest pains (especially those associated with coughing and weak hearts). These plants were said to stimulate sweating and to strengthen weak hearts. Some tribes took the tea to cure a dispirited heart. Powdered leaves were rubbed on the body to cool fevers. Blue giant-hyssop flowers were often included in Cree medicine bundles. More recently, teas made from the leaves and flowers of giant-hyssops have been used to relieve intestinal gas, to stimulate sweating and as a sedative to relieve tension.

OTHER USES: These plants are easily propogated from seeds, cuttings or root divisions. They grow best on limy, well-drained soil, preferably in full sun. Giant-hyssops make an attractive and useful addition to a wildflower garden.

DESCRIPTION: Large, perennial herbs with 4-sided stems 16–39" (40–100 cm) tall and paired, coarsely toothed leaves $1^1/_4$–4" (3–10 cm) long, smelling of anise. Flowers trumpet-shaped, 2-lipped, with a long stigma and 4 stamens projecting conspicuously from the mouth, whorled in spike-like clusters, in May to August. Fruits 4 small nutlets.

Blue giant-hyssop (both photos)

Nettle-leaved giant-hyssop (*A. urticifolia*) has rose or purplish to whitish flowers, $^3/_8$–$^5/_8$" (10–15 mm) long, and its leaves have long (to 2" [5 cm]) stalks and green, sparsely hairy to hairless lower surfaces. It grows in moist, open foothill, montane and subalpine sites from southern BC and Montana to Colorado.

Blue giant-hyssop (*A. foeniculum*) has blue flowers about $^3/_8$" (1 cm) long, and its leaves have shorter (less than $^5/_8$" [1.5 cm]) stalks and pale lower surfaces with a whitish bloom and dense, felt-like hairs. It grows in meadows and thickets in plains and foothills zones from BC and Alberta to Colorado.

Wild mints

Mentha spp.

Wild mint

FOOD: These plants can be eaten alone as greens, raw or cooked, but usually they are cooked with soups, stews and meats or used to flavor sauces, jellies and sweets. Mints make delicious, fragrant teas, cold drinks and even wine. Spearmint and peppermint have long been favorite flavors for gums, candies, syrups and liqueurs. All 3 species have been used to improve the flavor of other foods, including fruit juices, sauces and preserves, stout or overly yeasty beer, yogurt and fruit salads. Powdered mint leaves were sprinkled on berries and drying meat to repel insects, and dried mint plants were sometimes layered with stored, dried meat for flavor.

MEDICINE: The active medicinal ingredient, menthol (found in all 3 species), has been shown to expel gas from and relieve spasms of the digestive tract—hence the advent of the after-dinner mint. Menthol calms smooth muscles (eg., in the digestive tract), facilitates belching, and stimulates the liver to produce bile. It can help to prevent retching and vomiting and may also bring some relief from pain, coughing and sinus conjestion. Wild-mint tea was used by native peoples to treat colds, coughs, fevers, upset stomachs, gas, vomiting, kidney problems and headaches. Mint plants, menthol and peppermint oil have all been applied in poultices or lotions to relieve arthritis, tendinitis and rheumatism, and leaves were packed into and around aching teeth. Cold wild-mint tea was recommended for combating the effects of being struck by a whirlwind. In Europe, spearmint was an important medicine for treating diarrhea, stomach and bowel troubles, colic, gas, coughs, toothaches, headaches and hysteria. Peppermint tea may help to relieve morning sickness, but wild mint and spearmint should not be used for this purpose (see Warning). Peppermint extracts have been shown to be effective against some bacteria and viruses (including *Herpes simplex*).

OTHER USES: These aromatic plants were hung in dwellings as air-fresheners, and they were also crushed or chewed and rubbed on bodies as perfume (to improve one's love life). Mint was one of the aromatic plants boiled with traps to mask human scent. Mint oils from spearmint and peppermint have been used as a fragrance in toothpaste, soap and perfume and as a flavoring in toothpaste and mouthwash.

DESCRIPTION: Glandular-dotted perennials, smelling strongly of mint, with 4-sided stems bearing paired, sharp-toothed, short-stalked leaves $^3/_4$–3" (2–8 cm) long. Flowers light purple or pink to whitish, funnel-shaped, with 4 spreading lobes, whorled (sometimes forming spikes), in June to September. Fruits 4 oval nutlets.

The native species, **wild mint** or **field mint** (*M. arvensis*) produces whorls of flowers in its leaf axils. It grows in moist plains, foothill and montane sites from the Yukon and NWT to New Mexico.

Two European introductions have spread to moist sites on the plains and foothills from southern Canada or Montana to New Mexico. Both have spike-like flower clusters. **Spearmint** (*M. spicata*) has stalkless leaves and slender flower spikes ($^1/_4$–$^3/_8$" [5–10 mm] wide), whereas **peppermint** (*M. piperita*) has stalked leaves and thicker flower spikes ($^1/_4$–$^3/_8$" [10–15 mm] wide).

WARNING:
Wild mint and spearmint are high in pulegone, which stimulates the uterus, so they should not be used during pregnancy or heavy menstruation. Peppermint oil can cause heartburn and acid regurgitation when injested in large amounts, and it also causes rashes when applied to the skin. These plants should be used in moderation. As little as 1 tsp (5 ml) of pure menthol can be fatal. Also, some people are allergic to menthol. Use caution when giving infants or small children foods or medicines containing menthol or peppermint. They could gag, choke or even collapse from the intense fragrance.

Wild mint (both photos)

Wild bergamot

Monarda fistulosa

ALSO CALLED: horsemint
• *M. menthifolia.*

FOOD: Wild bergamot plants have been cooked alone as a potherb or used (like oregano) to flavor meat dishes, soups and stews. They can also make delicious, fragrant teas. This species smells much like oil-of-bergamot (extracted from the tropical tree, orange bergamot [*Citrus aurantium*]), which gives Earl Grey tea its distinctive flavor. Native peoples sprinkled the dried leaves on meat and drying berries to repel insects.

MEDICINE: The leaf tea has been used to treat coughs, colds, flu, fevers, pneumonia, insomnia, sore eyes, kidney and respiratory problems, nosebleeds, heart trouble, gas, cramps and indigestion and to expel intestinal worms. Studies have shown that wild bergamot stimulates sweating and helps to expel gas and that it contains the antiseptic 'thymol.' The leaves were packed around aching teeth or applied in poultices to relieve headaches and painful swollen joints and to heal rashes and acne. Dried, powdered plants (sometimes made into a paste) were rubbed on the head, face and limbs and inside the mouth to relieve fevers and headaches, or they were taken internally with water to treat fevers and sore throats. Dried leaves were sometimes wrapped around the neck in a cloth or strip of deerskin to relieve sore throats. Wild bergamot has been used to stimulate menstrual flow and to help expel afterbirth.

OTHER USES: Some tribes perfumed favorite horses with the chewed leaves of wild bergamot. Other people mixed bergamot and other sweet-smelling plants with 1–2 drops of beaver castor oil and a bit of water and applied it to their hair, body and clothing as perfume. Today, wild bergamot is used in potpourris and perfumes. Plants were burned on hot rocks in sweat baths as incense. Some tribes used wild bergamot as an insect repellent and burned it in smudges to drive insects away. This beautiful mint attracts many butterflies and hummingbirds, and it deserves a place in wildflower gardens. It is best grown from seed.

DESCRIPTION: Finely hairy, perennial herb, smelling strongly of mint, with square, unbranched stems 8–28" (20–70 cm) tall, bearing paired leaves 1–3" (2.5–8 cm) long. Flowers bright rose to purplish, 8–14" (20–35 mm) long, tubular, with a long, narrow upper lip and a 3-lobed lower lip, forming short, showy clusters, in June to August. Fruits 4 smooth nutlets. Wild bergamot grows in open, moist to moderately dry plains, foothill and montane sites from BC and Alberta to New Mexico.

Great mullein

Verbascum thapsus

ALSO CALLED: common mullein.

FOOD: Dried leaves have been steeped in hot water to make tea, but this tea should always be strained (see Warning).

MEDICINE: The leaves contain large amounts of mucilage, which soothes mucous membranes and has been shown to be anti-inflammatory. Mullein leaves and flowers were traditionally used to make medicinal teas for treating chest colds, asthma, bronchitis, coughs, kidney infections, diarrhea and dysentery. They were also applied as poultices to ulcers, tumors and hemorrhoids, or they were soaked in oil to make ear drops for curing earaches and killing ear mites. Chopped, dried leaves were smoked to relieve asthma, spasmodic coughing and fevers. Tea made with the stalks was used to treat cramps, fevers and migraine headaches. The roots are said to stimulate urination and to have an astringent effect on the urinary tract, so root tea was taken to tone the bladder, thereby reducing bed-wetting and incontinence. The flowers were soaked in oil to make anti-microbial drops for treating ear infections and removing warts. Mullein-flower tea was said to have a pain-killing, sedative effect.

OTHER USES: The large, fuzzy leaves provide soft toilet paper (but see Warning). They can also be used as padded insoles for foot-weary travelers. Some native peoples smoked the dried leaves (though too much was said to be poisonous) or used mullein smoke to clear the nostrils of horses with colds. The stems and leaves provided lamp wicks (before cotton was used), and dried flower stalks dipped in tallow were burned as torches. The seeds were put into ponds and slow streams as a narcotic fish poison, to stun fish. Roman women used the flowers to make yellow hair dye, and soap made with mullein ashes was said to return gray hair to its previous color.

DESCRIPTION: Grayish-felted, biennial herb with single stems 16–79" (40–200 cm) tall. Leaves many, 4–16" (10–40 cm) long, in basal rosettes (first and sometimes second year) alternate on the flowering stem (second year). Flowers bright yellow, $1/2$–$3/4$" (1–2 cm) wide, 5-lobed, forming dense, spike-like clusters 4–20" (10–50 cm) long, in June to August. Fruits round, wooly capsules, $1/4$–$3/8$" (7–10 mm) long. Great mullein is an introduced Eurasian weed that grows on disturbed ground in plains to subalpine zones from BC and Alberta to New Mexico.

WARNING:
This plant is generally considered safe for consumption in reasonable quantities, but it does contain tannin as well as rotenone and coumarin, which are classified as potentially dangerous by the US Food and Drug Administration. Mullein seeds are toxic. The leaf fuzz may irritate sensitive skin and throat membranes.

Common hound's-tongue

Cynoglossum officinale

FOOD: The young leaves have been boiled and eaten in small quantities, but this treatment is not recommended (see Warning).

MEDICINE: Because the fruits and leaves have a rough texture (like a dog's tongue), this plant was beaten with swine grease to make a salve for healing dog bites. Hound's-tongue is said to have soothing and mildly sedative properties, though there is no experimental evidence to substantiate these claims. The roots and leaves have been used in teas and decoctions to treat coughs, colds, shortness of breath, irritated membranes, diarrhea and dysentery. The leaves were boiled in wine to make a cure for dysentery and were applied as poultices to relieve insect bites, burns and hemorrhoids. These plants contain alantoin, a waxy compound that has been used to treat ulcers on the skin and in the intestine. They also contain heliosupine, an alkaloid that may be beneficial in the treatment for hemorrhoids.

OTHER USES: Perhaps because they are so hairy, the leaves were used in salves and ointments to cure baldness.

DESCRIPTION: Leafy, softly long-hairy, biennial herb, 12–32" (30–80 cm) tall, giving off an unpleasant smell when crushed. Leaves in a large basal rosette (first year), alternate on the stem (second year), 4–12" (10–30 cm) long. Flowers reddish-purple, 5-lobed, about ³/₈" (1 cm) across, borne on long, 1-sided, spreading branches, in May to July. Fruits prickly nutlets in ³/₈" (1 cm) wide clusters of 4. This Eurasian introduction is found on disturbed ground from BC and Alberta to New Mexico.

WARNING:
These plants contain alkaloids that depress the central nervous system and may cause liver damage and cancer. These compounds are most concentrated in roots and young leaves. In cattle and horses, common hound's-tongue causes disorders of the central nervous system, liver failure (in horses) and death. Horses also develop skin reactions following exposure to sunlight. In England, a family who ate these plants suffered vomiting, stupor and sleepiness, which continued for 40 hours, ending with 1 death. Handling these plants may cause skin reactions. Livestock become distressed when large numbers of nutlets adhere to their faces.

Gromwells

Lithospermum spp.

FOOD: Some tribes cooked and ate the large, deep taproots of yellow gromwell. Others gathered wayside gromwell seeds for food.

MEDICINE: These species were often used in similar ways as medicines. Teas made from the roots were taken to stop internal bleeding and improve appetite and were used as washes for treating skin and eye problems and rheumatism. Women drank the root tea every day as a contraceptive, and effects are said to have ranged from temporary to permanent sterility. Alcoholic extracts of these plants have been shown to eliminate the estrous cycle in laboratory mice. Powdered yellow gromwell roots were taken by people suffering from chest wounds. If a person had to stay awake, the plant was chewed and blown into his or her face and rubbed over the chest. Roots and seeds were ground together and steeped in hot water to make a medicinal tea that was cooled and used as an eyewash. Some people chewed the roots to relieve coughs and colds.

Wayside gromwell (both photos)

OTHER USES: The hard, shiny nutlets were used as decorative beads. The roots yielded a fast red dye, which was used in face and body paints and for coloring fabrics. Some tribes believed that wayside gromwell had magical powers. It was used by some as a charm to bring rain and by others to stop thunderstorms. Sometimes a plant was prayed over and then placed on an enemy's person, clothing or bedding to inflict sickness or bad luck.

DESCRIPTION: Clumped, hairy, perennial herbs with 8–24" (20–60 cm) tall stems bearing dark green, slender leaves $1^1/_4$–3" (3–8 cm) long. Flowers 5-lobed, clustered in the upper leaf axils, in April to July. Fruits 4 shiny, grayish-white to brownish nutlets.

Wayside gromwell, also called **wooly gromwell** or **lemonweed** (*L. ruderale*), has small ($^1/_8$–$^1/_4$" [4–6 mm] long, $^1/_4$–$^1/_2$" [6–13 mm] wide), pale yellow to greenish-tinted flowers and large ($^1/_8$–$^1/_4$" [4–6 mm]) nutlets. It grows on warm, dry, open slopes in plains, foothills and montane zones from southern BC and Alberta to Colorado.

Yellow gromwell or **narrow-leaved puccoon** (*L. incisum*) has larger ($^5/_8$–$1^1/_4$" [1.5–3 cm] long, $^1/_2$–$^3/_4$" [1–2 cm] wide), bright yellow to orange flowers and smaller ($^1/_8$" [3–4 mm]) nutlets. It grows on open, dry slopes in plains and foothills zones from BC and Alberta to New Mexico.

Bunchberry
Cornus canadensis

FOOD: The scarlet fruits are edible, but opinions of their flavor range from insipid to good. They were eaten raw as a trail nibble and were said to be good cooked in puddings. Bunchberries (often mixed with other fruits) have also been used whole to make sauces and preserves or cooked and strained to make syrups and jellies. Some people enjoy the crunchy, little, poppy-like seeds.

MEDICINE: Bunchberry is said to have anti-inflammatory, fever-reducing and pain-killing properties (rather like mild aspirin), without the stomach irritation and potential allergic effects of salicylates. It has been used to treat headaches, fevers, diarrhea, dysentery and inflammation of the stomach or large intestine. The berries were eaten and/or applied in poultices to reduce the potency of poisons. They were also chewed and applied to burns. Bunchberries were steeped in hot water to make a medicinal tea for treating paralysis, or they were boiled with tannin-rich plants (such as common bearberry [pp. 90–91] or commercial tea) to make a wash for relieving bee stings and poison-ivy rash. Native peoples used tea made with the entire plant to treat aches and pains, lung and kidney problems, coughs, fevers and fits. The root tea was given to colicky babies. Bunchberry has been studied as a potential anti-cancer agent.

DESCRIPTION: Perennial herb, 2–8" (5–20 cm) tall, with a whorl of 4–7 wintergreen, ³/₄–3" (2–8 cm) long leaves. Flowers tiny, in a dense clump at the center of 4 white to purple-tinged, petal-like bracts, forming single flower-like clusters about 1¹/₄" (3 cm) across, in May to August. Fruits bright red, berry-like drupes ¹/₄" (6–8 mm) wide, in dense clusters at the stem tips. Bunchberry grows in foothill and montane sites from Alaska to New Mexico.

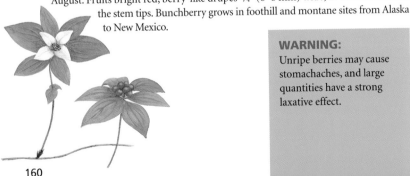

WARNING:
Unripe berries may cause stomachaches, and large quantities have a strong laxative effect.

160

Agoseris

Agoseris spp.

ALSO CALLED: false dandelion.

FOOD: The leaves have been eaten raw in salads or cooked as greens. The flowers are also edible raw and can be used to make tea.

MEDICINE: The Navajo used cold tea made from these plants as a lotion and drink to treat arrow and gunshot wounds and 'deer infection,' and to provide protection from witches. The wet leaves were rubbed on swollen arms, wrists and ankles.

OTHER USES: Agoseris sap (latex) contains rubbery compounds. The dried juice was collected from broken stems and leaves, rolled into little balls and used as gum. It was chewed for pleasure (like bubble gum) and to clean the teeth and then swallowed. Another name for these plants was 'Indian bubble gum.' The leaves were also dried and then chewed to extract the gum.

DESCRIPTION: Perennial herbs with milky sap and slender basal leaves 2–14" (5–35 cm) long. Flowerheads with ray florets and several overlapping rows of slender involucral bracts, borne singly on leafless stalks 4–28" (10–70 cm) tall, in May to September. Fruits smooth seed-like achenes $1/4$–$3/8$" (5–9 mm) long, tipped with a starburst of white hairs on a stalk-like beak.

Orange agoseris (*A. aurantiaca*) has burnt-orange flowers, and its achenes have slender beaks more than $1/2$ as long as the body. It grows on open foothill, montane, subalpine and alpine sites from BC to Alberta to New Mexico.

Short-beaked agoseris (*A. glauca*) has yellow flowers, and its achenes have ribbed beaks less than $1/2$ as long as the body. It grows in open, grassy foothill and montane slopes from BC and Alberta to New Mexico.

Short-beaked agoseris (top)
Orange agoseris (middle)
Agoseris seed head (bottom)

161

Common dandelion
Taraxacum officinale

FOOD: All parts of common dandelion are edible. Young leaves have been used in salads (fresh or cooked and chilled) or served hot as a cooked vegetable—scalloped, baked or added to meat dishes and soups. They can also be mixed in a blender with tomato juice, Worchestershire sauce and Tabasco to make a vitamin-rich cold drink. Older leaves or leaves growing in sunny areas are especially bitter. Bitterness can be reduced by growing plants away from light (under straw or baskets), by removing the leaf midveins, and by boiling the plants in at least 2 changes of water. The roots, dug in spring or autumn, peeled, sliced and cooked in 2 changes of water with a pinch of baking soda, have been used as a cooked vegetable, similar to parsnips. Dandelion roots can also be used raw in salads or can be dried, roasted slowly until dark brown throughout, and ground to make a coffee substitute (though generally considered inferior to coffee made with chicory, p. 174). The flower petals produce a highly esteemed, delicately flavored, pale yellow wine. They can also make a pretty addition to pancakes. Unopened buds can be eaten raw in salads, cooked in pancakes and fritters or pickled. The chewy seeds (without their fluffy parachutes) have been eaten as a nibble, ground into flour or used to grow sprouts.

MEDICINE: Dandelions are rich in vitamins A, C, E and B-complex, iron, calcium and potassium. Dandelion root or leaf tea was recommended as a mild laxative that would also stimulate urination, salivation and the secretion of gastric juices and bile, improve appetite and generally tone the whole system. It has been used to treat liver, urinary tract and digestive problems. The roots have been reported to lower blood sugar and cholesterol levels, to lower blood pressure, to reduce inflammation, to have anti-microbial effects (against *Candida albicans* in particular) and to aid weight loss. They also contain the sugar inulin, which is said to be an immune-system stimulant. Historically, dandelion flowers were used to treat jaundice and other liver ailments, perhaps because of their yellow color. Recently, they have been shown to contain large amounts of lecithin, which has been shown to prevent cirrhosis in chimpanzees. The milky juice, applied 3 times daily for 7–10 days, was said to kill warts.

WARNING:
Never collect dandelions that might have been sprayed with herbicides. The milky sap can cause rashes on sensitive skin.

OTHER USES: Dandelion leaves have been used as a substitute for mulberry leaves for feeding silk worms. The flowers produce a yellow dye, and the roots give a magenta color.

DESCRIPTION: Perennial herb with bitter, milky sap and thick taproots. Leaves basal, 2–16" (5–40 cm) long, with triangular, backward-pointing lobes. Flowerheads bright yellow, 1³/₈–2" (3.5–5 cm) across, with ray florets only, borne on hollow, leafless stalks, in May to August. Fruits yellowish to pale gray or olive-brown, spiny-ribbed seed-like achenes ¹/₈" (3–4 mm) long, tipped with a stalked cluster of white hairs, forming fluffy, round heads. This widespread European introduction grows on disturbed, cultivated and waste ground from Alaska to New Mexico.

Salsifies

Tragopogon spp.

FOOD: The thick, fleshy roots of all 3 species, collected before the flower stalks appear, can be eaten raw, roasted, fried or boiled, but those of the cultivated species (common salsify) are the largest and tastiest. They are said to taste like parsnips. Salsify roots have also been dried, ground and added to cakes, or roasted until they are dark brown, ground and used as a coffee substitute. Tender young leaves, buds and flowers have been added to salads or served as a cooked vegetable. The young stalks and root crowns can be gently simmered, like asparagus (*Asparagus officinalis*) and artichokes (*Cynara scolymus*), respectively. The seeds can be used to make tasty sprouts.

MEDICINE: These plants were used for many years to relieve heartburn and to stimulate urination (especially in the treatment of kidney stones). The milky juice was taken to cure indigestion and was applied to dressings to stop oozing and bleeding of sores and wounds. Common salsify tea was used as a drink or lotion to treat the bites of mad coyotes on both humans and livestock.

OTHER USES: Some tribes gathered the rubbery sap from broken stems and leaves, dried it and rolled it into balls. It was chewed like gum.

DESCRIPTION: Mostly biennial, 8–40" (20–100 cm) tall herbs with milky sap. Stems erect, bearing many slender, grass-like leaves 2–12" (5–30 cm) long with clasping bases. Flowerheads about 2–2¹/₂" (5–6 cm) across, with ray florets only and a single row of involucral bracts, borne singly, in April to August. Fruits slender, spiny seed-like achenes, each with a slender, stalk-like beak bearing a tuft of feathery bristles, forming dandelion-like seed heads up to 4" (10 cm) wide.

Yellow salsify (both images)

Meadow salsify or **meadow goat's-beard** (*T. pratensis*) has slender upper stalks, small (5/8–1" [15–25 mm] long) achenes and about 8 involucral bracts, which are shorter than the yellow florets.

Yellow salsify or **common goat's-beard** (*T. dubius*) has enlarged upper stalks, large (1–1½" [25–40 mm]) achenes and about 13 involucral bracts, which are longer than the yellow florets.

Common salsify or **oyster plant** (*T. porrifolius*) has enlarged upper stalks, large (1–1½" [25–40 mm]) achenes and about 8 involucral bracts, which are longer than the blue florets.

All 3 species are European weeds that grow on disturbed sites across North America.

Yellow salsify (top); Common salsify (above)

Sow thistles

Sonchus spp.

FOOD: Young leaves have been added to salads or cooked as a hot vegetable (served with butter and seasonings or vinegar) or as a potherb. They are sometimes added directly to curries and rice dishes, but they are usually quite bitter and require at least 1 change of water. Older leaves soon become very bitter and tough.

MEDICINE: In the 13th century, sow thistle was recommended to prolong the virility of gentlemen, perhaps because of its milky sap. It was also believed to increase the flow of milk in nursing mothers. Some herbalists recommend sow thistle for use in salves for healing hemorrhoids, ulcers and other skin irritations and as a bitter and a tonic. The dried sap of annual sow thistle was used to treat opium addiction. Medicinal teas made from perennial sow thistle have been used to calm nerves and to treat asthma, bronchitis and coughs. The leaves were used in poultices and washes for relieving swelling and inflammation, and the milky juice was used to treat inflamed eyes.

OTHER USES: The white, milky sap was believed to clear the complexion. It was also dried in balls and chewed like gum. The flowers can be used as a pleasant-tasting, gummy chew. Sow thistles are sometimes given to animals that have lost interest in other forage. Pigs are especially fond of these plants—hence the name sow thistle—but older European lore often associated sow thistles with hares, which were said to hide among the plants and to eat the leaves to cool their blood after being pursued.

Prickly sow thistle (top); Perennial sow thistle (above)

DESCRIPTION: Tall, often bluish-green herbs with milky juice. Leaves usually prickly-edged with bases clasping the stem. Flowerheads yellow, with ray florets only, forming branched clusters, in July to October. Fruits slightly flattened, ribbed seed-like achenes, each tipped with a tuft of white, hair-like bristles.

Perennial sow thistle (*S. arvensis*) has deep, creeping rootstocks, 1¼–2" (3–5 cm) wide flowerheads and glandular-hairy involucral bracts.

Two annual sow thistles also grow in the Rockies.

Prickly sow thistle (*S. asper*) has stiff, spiny, usually unlobed leaves, with rounded basal lobes that clasp the stem.

Annual sow thistle (*S. oleraceus*) has soft, non-spiny, rather dandelion-like leaves, with sharply pointed clasping lobes.

All 3 species were introduced from Europe and grow on disturbed sites from BC and Alberta to New Mexico.

Perennial sow thistle (both images)

Thistles

Cirsium spp.

Creeping thistle

FOOD: Flavor and texture of thistles varies with species, age and habitat, ranging from tender and sweet to tough and bitter. The taproots of young biennial plants (at the end of the first summer and before bolting the following spring) could be eaten raw, but usually they were roasted for several hours in fire pits or were boiled. Cooked roots were sliced and then fried or mashed, dried and ground into flour. Because they contain a relatively indigestible sugar, inulin, raw thistle roots often cause gas. Some roots turn sugary when roasted. The stems and leaves are often sweet and juicy, but before they can be eaten, they have to be peeled to remove their prickles—hold the plant upside-down and peel from the bottom to the top with a sharp knife. Peeled parts can then be eaten raw or cooked as a vegetable. The immature flowerheads can be eaten raw or steamed and served with lemon butter. Some say they are better than their well-known relative, the artichoke (*Cynara scolymus*). Because these nutritious plants are so widespread and easily identified, they are an excellent survival food.

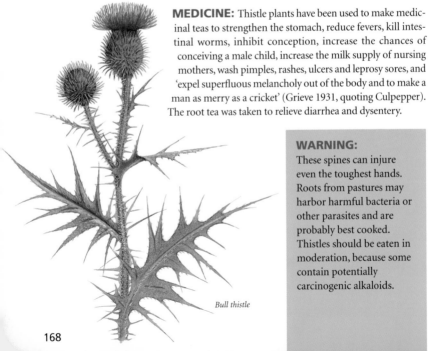

MEDICINE: Thistle plants have been used to make medicinal teas to strengthen the stomach, reduce fevers, kill intestinal worms, inhibit conception, increase the chances of conceiving a male child, increase the milk supply of nursing mothers, wash pimples, rashes, ulcers and leprosy sores, and 'expel superfluous melancholy out of the body and to make a man as merry as a cricket' (Grieve 1931, quoting Culpepper). The root tea was taken to relieve diarrhea and dysentery.

WARNING: These spines can injure even the toughest hands. Roots from pastures may harbor harmful bacteria or other parasites and are probably best cooked. Thistles should be eaten in moderation, because some contain potentially carcinogenic alkaloids.

Bull thistle

OTHER USES: The seed fluff makes good tinder, pillow stuffing and insulation. Thistle flower petals provide a pleasant substitute for chewing gum. Tough, fibrous older stems can be twisted to make strong twine. Thistle-seed oil was used as lamp oil in Europe.

DESCRIPTION: Erect, prickly herbs with alternate, spiny-toothed to deeply lobed leaves. Flowerheads with tubular (disc) florets above overlapping rows of involucral bracts, few to many, in head-like or spreading clusters. Fruits hairless, ribbed seed-like achenes, tipped with feathery bristles.

Hooker's thistle

Creeping thistle or **Canada thistle** (*C. arvense*) is a perennial with deep, spreading rootstocks. Unlike the other thistles, it has small (1/2–1" [12–25 mm] wide), pinkish-purple flowerheads that are either male or female.

Bull thistle (*C. vulgare*) is easily recognized by its wide (about 2" [5 cm]), showy, rose-purple flowerheads and the rough-hairy upper surface of its leaves.

Both of these weedy thistles were introduced from Europe and grow on disturbed sites from BC and Alberta to New Mexico.

Two common native species (below) have cobwebby-wooly plants and short-stalked, creamy to dirty white flowerheads.

Hooker's thistle (*C. hookerianum*) has relatively broad involucral bracts, often with enlarged, fringed tips. It grows in moist montane and subalpine sites from BC and Alberta to Idaho and Montana.

Leafy thistle (*C. foliosum*) has hairy, slender, tapered involucral bracts. It grows on moist sites in foothills to subalpine zones from the southern Yukon or NWT to Wyoming.

Bull thistle

Burdocks

Arctium spp.

FOOD: These large, vitamin- and iron-rich plants were originally brought to North America as food plants. All parts of the plants are edible. Young leaves have been used in salads or boiled as a potherb, though in order to break down the tough fibers the leaves may require 1–2 changes of boiling water and the addition of baking soda. They are said to be excellent in soups and stews. The roots of first-year plants have been peeled, diced or sliced and used in stir-fries and soups or served as a hot vegetable. Mashed roots can also be formed into patties and fried. Some tribes dried burdock roots for winter supplies or roasted and ground them for use as a coffee substitute. The white pith of young leaf and flower stalks has been added to salads or cooked as a vegetable. It was also simmered in syrup to make candy or soaked in vinegar and seasonings to make pickles.

MEDICINE: Burdocks were widely used in tonics for purifying the blood, and they are still recommended as a safe but powerful liver tonic. For centuries burdock was used in medicinal teas for treating gout, liver and kidney problems, rheumatism, vertigo, high blood pressure, measles and gonorrhea. The leaves provided poultices for healing burns, ulcers and sores, as well as teas and washes for treating hair loss, hives, eczema, psoriasis and skin infections. The seeds were also used in washes and poultices to treat bruises, abscesses, insect

WARNING:
Pregnant women and people with diabetes should not use burdock. In women it can cause spotting and even miscarriage. It has also shown hypoglycemic activity. People with sensitive skin may develop rashes from contact with these plants.

Common burdock (both photos)

bites, snake bites, scarlet fever and smallpox. They are also said to stimulate urination and have been used to treat water retention and high uric acid levels. The roots were boiled to make an antidote for use after eating poisonous food, especially poisonous mushrooms. In Japan, common burdock is popular because of its reputation for increasing endurance and sexual virility. Studies have shown that burdock can reduce blood-sugar levels, dissolve bladder stones, and stimulate liver and bile function. It may also inhibit mutations, slow tumour growth and relieve water retention, rheumatism, skin problems associated with liver dysfunction and high blood pressure.

OTHER USES: The hooked bristles of these bur-like flowerheads are said to have inspired the invention of Velcro.

DESCRIPTION: Robust, biennial herbs with mostly heart-shaped leaves that are thinly white-wooly beneath. Flowerheads bur-like, with slender, tubular, pink to purplish florets above overlapping rows of hook-tipped involucral bracts, borne in branched clusters, in August to October. Fruits 3–5-sided, hairless seed-like achenes.

Great burdock (*A. lappa*) has open clusters of large (1–1⅝" [2.5–4 cm] wide), long-stalked flowerheads, whereas **common burdock** (*A. minus*) has smaller (⅝–1" [1.5–2.5 cm] wide), stalkless flowerheads that are scattered in elongated groups. **Wooly burdock** (*A. tomentosum*) is easily recognized by its cobwebby-wooly flowerheads.

All three species were introduced from Europe. Great burdock and common burdock now grow on disturbed sites in BC, Alberta and the northern US. Common burdock also reaches south to Colorado. Wooly burdock grows on disturbed sites in the US only.

Burdocks could be confused with **common cocklebur** (*Xanthium strumarium*), which is reported to have caused poisoning of cattle, sheep, horses and swine. Cockleburs are easily distinguished by their flowerheads, in which the hooked bracts are fused into solid, nut-like burs, and by their leaves, which are rough rather than velvety.

Wooly burdock (both photos)

171

Wild sages

Artemisia spp.

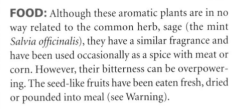

FOOD: Although these aromatic plants are in no way related to the common herb, sage (the mint *Salvia officinalis*), they have a similar fragrance and have been used occasionally as a spice with meat or corn. However, their bitterness can be overpowering. The seed-like fruits have been eaten fresh, dried or pounded into meal (see Warning).

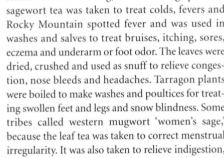

MEDICINE: Strong, bitter-tasting pasture-sagewort tea was taken to treat colds, fevers and Rocky Mountain spotted fever and was used in washes and salves to treat bruises, itching, sores, eczema and underarm or foot odor. The leaves were dried, crushed and used as snuff to relieve congestion, nose bleeds and headaches. Tarragon plants were boiled to make washes and poultices for treating swollen feet and legs and snow blindness. Some tribes called western mugwort 'women's sage,' because the leaf tea was taken to correct menstrual

Pasture sagewort

irregularity. It was also taken to relieve indigestion, coughs and chest infections. Western mugwort smoke was used to disinfect contaminated areas and to revive patients from comas. Northern-wormwood tea was taken to relieve difficulties with urination or bowel movements, to ease delivery of babies and to cause abortions. As the name suggests, it was also used to expel worms—especially pinworms and roundworms. Chewed or powdered sage leaves were often applied to cuts, sores and blisters, and they have been shown to have anti-bacterial properties.

OTHER USES: Sage plants and smudges repel insects, and these aromatic herbs were also believed to drive away evil. They were widely used in religious ceremonies, both in smoking mixtures and as incense for purifying implements and people. Western mugwort was sometimes spread along the edges of ceremonial lodges and made into protective wreaths and bracelets. If someone committed a taboo, he or she was purified by a whipping with a bundle of western mugwort. The pulverized roots of northern wormwood were used as perfume. It was believed that if you placed them on the face of a sleeping man, he would not wake up and you could then steal his horses. The soft leaves of western mugwort provided toilet paper, sanitary napkins and deodorizing shoe liners, and the leaf tea was used as a hair tonic.

DESCRIPTION: Aromatic, perennial herbs with alternate leaves. Flowerheads small, yellowish, with disc florets only, borne in narrow to open branched clusters, in June to October. Fruits hairless seed-like achenes.

WARNING:
Most sages can cause allergic reactions and many are considered toxic, so these plants should always be used with caution. Large doses of tarragon's essential oil caused cancer in experimental animals.

Tarragon (*A. dracunculus*) is 1¹/₂–3¹/₂' (50–100 cm) tall, with essentially hairless, linear (rarely lobed) leaves and ¹/₈" (2–4 mm) wide flowerheads in open clusters. It grows on dry, open sites in plains, foothills and montane zones from the southern Yukon to New Mexico.

Western mugwort (*A. ludoviciana*) also has toothless leaves, but they are densely wooly on both sides. It grows in dry, open plains, foothill and montane sites from BC and Alberta to New Mexico.

Pasture sagewort or **prairie sagewort** (*A. frigida*) is 4–16" (10–40 cm) tall and has densely wooly, feathery leaves that are finely divided 2–3 times into linear segments up to ¹/₁₆" (1 mm) wide. It grows on dry plains to subalpine sites from Alaska to New Mexico.

Northern wormwood or **plains wormwood** (*A. campestris*) is larger (1–2' [30–60 cm] tall) than western mugwort, gray-green (hairy, but not silvery-wooly) and less strongly scented. It grows on dry slopes in plains, foothills and montane zones from Alaska to New Mexico.

Northern wormwood

Western mugwort

*C*hicory

Cichorium intybus

ALSO CALLED: blue sailors.

FOOD: The leaves can be eaten raw in salads, but become very bitter with age and with exposure to sunlight. Green leaves usually require at least 1 change of water during cooking. Chicory is sometimes sold as a tangy (though bitter) flavoring for soups and stews. Young, white, underground parts or young plants grown in darkness are best. Young plants make an excellent cooked vegetable. Belgian endive is a variety of *C. intybus*. Chicory roots can be eaten raw, boiled or roasted. They are said to have a carrot-like flavor when young. More often, however, they are split, dried, roasted until brown throughout and ground to make a coffee substitute. Use about 1¹/₂ tsp (8 ml) per cup; if it is mixed with coffee, use 2 parts coffee to 1 part chicory and reduce the total grounds by ¹/₃. The roots are best collected before or well after the plants have flowered. The flowerheads have been added to salads, pickled or cooked in soups and stews. Chicory is grown commerically as a source of fructose and maltol (a sugar enhancer).

MEDICINE: These plants are rich in vitamins A and C. Chicory-root tea/coffee is reputed to improve appetite and stimulate bile secretion and urination. Historically, chicory-root tea was used to treat liver problems, gout, skin infections, rheumatism, fevers, inflammations, nausea, lung problems, typhoid and cancer. Mashed roots were used in poultices to heal sores from fevers and venereal disease. Studies have shown chicory root extracts to lower blood sugar and to have slightly sedative, mildly laxative, anti-bacterial, anti-mutagenic, liver-protective and anti-inflammatory effects. Chicory slowed and weakened the pulse of test animals, so it has been suggested for study in the treatment of heart problems. The flowers were used in eyewashes for treating inflamed eyelids.

DESCRIPTION: Perennial herb, 1–5' (30–150 cm) tall, with spreading branches, milky sap and deep taproots. Leaves mainly basal, dandelion-like, 3–8" (8–20 cm) long. Flowerheads blue (rarely white), about (3–4 cm) wide, with ray florets only and 2 rows of involucral bracts, essentially stalkless, single or in 2s or 3s, in July to October. Fruits hairless, 5-sided seed-like achenes ¹/₈" (2–3 mm) long, tipped with tiny scales. This introduced weed from Eurasia grows on disturbed ground in plains, foothills and montane zones from BC and Alberta to New Mexico.

WARNING:
Excessive and/or prolonged use of chicory may cause sluggish digestion and damage to the retinas. Some people develop rashes from contact with chicory coffee. People with gall stones should consult a doctor before using chicory.

Yarrow

Achillea millefolium

FOOD: Some sources suggest parboiled yarrow as a vegetable, but most consider it too bitter to eat. In Sweden, these plants sometimes replaced hops in beer.

MEDICINE: Yarrow has been used for thousands of years as a styptic—a plant that stops bleeding. Achilles, the Greek hero for whom this genus was named, was said to have saved the lives of many soldiers by applying yarrow to their wounds. These plants contain alkaloids that have been shown to reduce clotting time and have been used to suppress menstruation. They also have sedative, pain-killing, antiseptic, anti-inflammatory and antispasmodic constituents that may help to relieve menstrual cramps. Yarrow leaves have been used in washes, salves and poultices for treating burns, boils, open sores, pimples, mosquito bites, earaches, sore eyes and aching backs and legs. The tea has been taken as a tonic and as a treatment for colds and fevers, because it stimulates sweating and lowers blood pressure. These plants also contain substances that stimulate salivation and the secretion of bile and gastric juices. Yarrow has been used to improve appetite and digestion, to speed labor and heal the uterus after birth, and to treat diarrhea, urinary tract infections and even diabetes. Mashed leaves or roots were used as a topical anesthetic on aching teeth.

OTHER USES: Dried yarrow has been used for perfume and bath powder. Fresh leaves can be rubbed on the skin as an effective (though temporary) insect repellent. These hardy, attractive plants are excellent in wildflower gardens, but their spreading rootstocks often creep into areas where they are not wanted. The attractive, spicy-scented dried flowers are lovely in dried flower arrangements. Yarrow tea is said to make an excellent hair rinse.

DESCRIPTION: Aromatic, perennial herb, 4–32" (10–80 cm) tall, from spreading rootstocks. Leaves alternate, fern-like, 2–3-times pinnately divided into fine segments 1/16" (1–2 mm) wide. Flowerheads white (sometimes pinkish), about 1/4" (5 mm) across, with about 5 white ray florets around 10–30 yellowish disc florets, forming flat-topped clusters 3/4–4" (2–10 cm) across, in May to September. Fruits hairless, flattened seed-like achenes. Yarrow grows in open, often disturbed sites in plains to alpine zones from Alaska to New Mexico.

WARNING:
People with sensitive skin may react to these plants. Yarrow contains thujone, which is toxic in large doses and can cause miscarriages.

Pearly everlasting

Anaphalis margaritacea

FOOD: The leaves and young plants have been used as a potherb.

MEDICINE: Pearly everlasting has been reported to have anti-inflammatory and astringent effects and to be a mild antihistamine, expectorant, sedative and diaphoretic (increasing perspiration). It has been used in medicinal teas and gargles to treat swollen mucous membranes associated with colds, bronchial coughs, asthma, throat infections, upset stomachs, diarrhea, bleeding of the bowels and dysentery. Native Americans smoked the dried plants to relieve headaches and throat or lung problems, and they used a smudge of pearly everlasting (either alone or mixed with mint) to treat paralysis. Sometimes plants were boiled in water and used to steam rheumatic joints, or they were applied in poultices to rheumatic joints, sores, bruises and swellings. Leaf and flower poultices have been recommended for healing sunburn and moderate burns from heat and friction. Juice from fresh plants was said to be an aphrodisiac.

OTHER USES: Several tribes smoked pearly everlasting, either alone or mixed with other plants such as tobacco (*Nicotiana tabacum*). The fuzziest leaves were said to be best. Men chewed the plants and rubbed the paste on their bodies to gain strength, energy and protection from danger. Pearly everlasting smoke was used to purify gifts being left for the spirits and to protect houses from witches. Similarly, a smudge of pearly everlasting and beaver gall bladders was said to revive people who had fainted and to ward off troublesome ghosts after someone in the family died. The flower stalks have a pleasant fragrance, and they also keep their shape and color when dried, making pearly everlasting an excellent subject for dried flower arrangements.

DESCRIPTION: White-wooly, perennial herb with leafy stems about 8–24" (20–60 cm) tall. Leaves alternate, slender, 1¼–4" (3–10 cm) long, with a conspicuous midvein and down-rolled edges. Basal leaves small, soon withered. Flowerheads white, up to ³⁄₈" (1 cm) across, with showy, papery involucral bracts around smaller, yellow to brownish centers of disc florets, forming flat-topped clusters, in July to September. Fruits roughened seed-like achenes tipped with short, white hairs. Pearly everlasting grows on open, often disturbed sites in foothill, montane and subalpine zones from BC and Alberta to New Mexico.

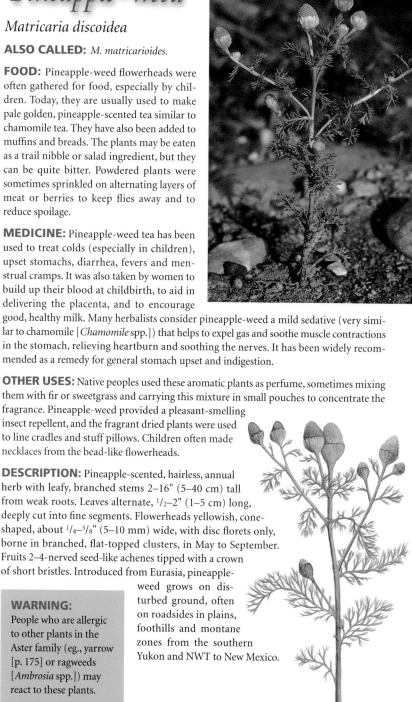

Pineapple-weed

Matricaria discoidea

ALSO CALLED: *M. matricarioides.*

FOOD: Pineapple-weed flowerheads were often gathered for food, especially by children. Today, they are usually used to make pale golden, pineapple-scented tea similar to chamomile tea. They have also been added to muffins and breads. The plants may be eaten as a trail nibble or salad ingredient, but they can be quite bitter. Powdered plants were sometimes sprinkled on alternating layers of meat or berries to keep flies away and to reduce spoilage.

MEDICINE: Pineapple-weed tea has been used to treat colds (especially in children), upset stomachs, diarrhea, fevers and menstrual cramps. It was also taken by women to build up their blood at childbirth, to aid in delivering the placenta, and to encourage good, healthy milk. Many herbalists consider pineapple-weed a mild sedative (very similar to chamomile [*Chamomile* spp.]) that helps to expel gas and soothe muscle contractions in the stomach, relieving heartburn and soothing the nerves. It has been widely recommended as a remedy for general stomach upset and indigestion.

OTHER USES: Native peoples used these aromatic plants as perfume, sometimes mixing them with fir or sweetgrass and carrying this mixture in small pouches to concentrate the fragrance. Pineapple-weed provided a pleasant-smelling insect repellent, and the fragrant dried plants were used to line cradles and stuff pillows. Children often made necklaces from the bead-like flowerheads.

DESCRIPTION: Pineapple-scented, hairless, annual herb with leafy, branched stems 2–16" (5–40 cm) tall from weak roots. Leaves alternate, $1/2$–2" (1–5 cm) long, deeply cut into fine segments. Flowerheads yellowish, cone-shaped, about $1/4$–$3/8$" (5–10 mm) wide, with disc florets only, borne in branched, flat-topped clusters, in May to September. Fruits 2–4-nerved seed-like achenes tipped with a crown of short bristles. Introduced from Eurasia, pineapple-weed grows on disturbed ground, often on roadsides in plains, foothills and montane zones from the southern Yukon and NWT to New Mexico.

WARNING:
People who are allergic to other plants in the Aster family (eg., yarrow [p. 175] or ragweeds [*Ambrosia* spp.]) may react to these plants.

Common tansy
Tanacetum vulgare

FOOD: Sweet cakes with snippets of tansy leaves were once popular in Europe as a spring tonic, but they are seldom made today. In small amounts, tansy has been used to replace sage, but this substitution is not recommended (see Warning).

MEDICINE: Tansy tea was used to treat many problems, including irregular menstruation, rheumatism, gout, jaundice, indigestion, lack of appetite, nerve pain, migraine headaches and hysteria. The oil or strong tea was used to induce abortions, sometimes with fatal results for the woman. Tansy seed was taken to expel worms, to soothe nervousness and enhance sleep, and to treat bladder problems and gout. It was also used in liniments and cosmetic lotions. Leaf poultices were applied to cuts and bruises. Common tansy contains compounds called 'parthenolides,' which may help to prevent migraine headaches. Some studies suggest that tansy may relieve intestinal spasms, fight intestinal worms, stimulate bile activity, reduce blood-lipid levels, influence blood-sugar levels, kill bacteria and fungi, combat tumors and increase resistance to the encephalitis virus. However, it is too toxic to warrant use.

OTHER USES: Tansy was strewn on floors, under mattresses, between blankets and clothing and around food to kill and repel insects. Strong tansy tea was recommended as a garden insecticide and as a wash for treating scabies or lice. Plants were rubbed on corpses and placed in coffins to mask odors and prevent decay. The bright yellow flower clusters are popular in dried flower arrangements. Some sources recommend tansy as a long-lived garden flower, but it spreads rapidly and is very difficult to contain. Also, in many areas it is classified as a noxious weed.

DESCRIPTION: Aromatic, perennial herb, dotted with pitted glands. Stems leafy, erect, branched, about $1^{1}/_{2}$–$3^{1}/_{4}$' (50–100 cm) tall, from spreading rootstocks. Leaves alternate, fern-like, 4–8" (10–20 cm) long, deeply cut into slender, sharp-toothed or lobed segments. Flowerheads deep yellow, button-like clusters of disc florets about $^{1}/_{4}$–$^{3}/_{8}$" (5–10 mm) across, forming flat-topped clusters, in July to September. Fruits glandular-dotted seed-like achenes. Common tansy is a European introduction that now grows on disturbed sites at lower elevations from BC and Alberta to New Mexico.

WARNING:
The pungent, volatile oils of these plants are poisonous—a small quantity can kill in 2–4 hours. The US Federal Department of Agriculture prohibits the sale of tansy as a food or medicine. People have died from taking tansy oil or drinking tansy tea. People with sensitive skin may develop rashes from contact with tansy.

Tall coneflower & Black-eyed Susan

Rudbeckia spp.

Black-eyed Susan

FOOD: These plants are often bitter and astringent, but young shoots of tall coneflower were eaten for good health. The hairier black-eyed Susan was usually used as medicine.

MEDICINE: Tall coneflower was generally considered a kidney medicine, and it was sold commercially for use as a tonic, diuretic (for stimulating urination) and soothing balm, especially recommended for long-term inflammation of the mucous membranes of the urinary tract. Its roots were also mixed with other herbs in teas for relieving indigestion, and its flowers were included with other plants in poultices for treating burns. Black-eyed Susan was also said to increase urination and to have a mild, stimulating effect on the heart. Root tea was taken to treat colds and to expel worms. It was also used as a wash to heal sores, snakebites and swellings. Juice from the root was dropped into the ear to cure earaches. Recent studies report that coneflower root extracts can be more effective at stimulating the immune system than extracts of echinacea (*Echinacea* spp.), so these plants are being studied as a potential treatment for AIDS.

OTHER USES: Black-eyed Susan was used in colonial times for treating saddle sores on horses. These showy, hardy wildflowers make an attractive addition to sunny gardens.

DESCRIPTION: Biennial or perennial herbs with alternate leaves on erect stems from spreading rootstocks. Flowerheads usually showy, with few (6–20) ray florets around a $1/2$–$3/4$" (1–2 cm) wide, hemispheric to cone-shaped cluster of disc florets, long-stalked, 1 to few, in June to August. Fruits hairless, 4-sided seed-like achenes.

WARNING:
Some people have an allergic reaction to these plants. Black-eyed-Susan tea should be strained to remove the irritating hairs. Poisoning and death of pigs, sheep and horses have been reported following ingestion of tall coneflower.

Black-eyed Susan (*R. hirta*) has rough-hairy, undivided (sometimes toothed) leaves and bright yellow to orange flowerheads with dark, purplish-brown centers. It grows on open, often disturbed ground in plains, foothills and montane zones from BC and Alberta to Colorado.

Tall coneflower (*R. laciniata*) has hairless, deeply cut leaves, and its yellow flowerheads have yellow or grayish centers. It grows on moist sites in plains, foothills and montane zones from Montana to New Mexico.

Common sunflower

Helianthus annuus

FOOD: This plant was one of the few cultivated by tribes on the plains. Some seeds were eaten raw, but most were dried, parched, ground lightly to break their shells, and poured into a container of water, where the kernels sank and the shells floated. The shells were skimmed off, roasted, and used to make a coffee-like beverage. The kernels were eaten whole or ground into meal, which was boiled in water to make gruel or mixed with bone marrow or grease to make cakes. Small, flat sunflower-seed cakes provided light-weight, high-energy food on journeys. Crushed seeds were boiled in water, and the oil was skimmed from the surface and used like olive oil. Sunflower seed sprouts are also edible.

MEDICINE: Common sunflower has been used to treat many ailments, including rheumatism, headaches, kidney weakness, malaria, worms, fatigue, sores, swellings, blisters and even prenatal problems believed to have been caused by an eclipse of the sun. Experimentally, it has been shown to have hypo-glycemic activity. Sunflower seed oil was sometimes taken to relieve coughs and laryngitis. The flowers were used to make medicinal teas for treating lung problems, and the leaf tea was given to relieve high fevers. Sunflower leaves were applied as poultices to snakebites and spider bites. To remove a wart, the wart was scratched, and the pith from a sunflower stalk was then burned on top of it.

OTHER USES: Oil from the seeds was rubbed on the body and used to groom hair. It has also been used in candle making and soap making. The flowers provided a superior yellow dye, and the seeds were used to make black or purple dyes. Sunflower stem fibers were made into paper. The sunflower is the state flower of Kansas.

DESCRIPTION: Coarse, rough-hairy, annual herb, 1¹/₄–6¹/₂' (40–200 cm) tall. Leaves mainly alternate (lowest leaves opposite), stalked, oval to heart-shaped, 4–8" (10–20 cm) long. Flowerheads 2–4" (5–10 cm) wide, with bright yellow ray florets around a large (over ³/₄" ([2 cm])across, reddish-brown (rarely yellow) button of disc florets, borne singly or in open, few-flowered clusters, in June to September. Fruits thick, 2–4-sided seed-like achenes ¹/₄–³/₈" (5–10 mm) long (sunflower seeds). Common sunflower grows on open, often disturbed sites in plains and foothills zones from BC and Alberta to New Mexico.

WARNING:
Some people have allergic reactions to sunflower pollen and/or sunflower plant extracts.

Arrow-leaved balsamroot

Balsamorhiza sagittata

FOOD: Young leaves and shoots (often dug before they had emerged from the ground) were eaten raw, boiled or steamed. Immature flower stems and young leaf stalks were peeled and eaten raw, like celery, or were cooked. The peeled roots have been eaten raw, but their bitter, strongly pine-scented sap contains a complex, hard-to-digest sugar (inulin). This sugar is converted to sweet-tasting, easily-digested fructose by slow cooking, so roots were usually roasted or steamed for several hours or days. Dried, cooked roots were stored and then soaked overnight prior to use. The small, sunflower-like seeds were dried or roasted and pounded into meal. This meal was mixed with grease and formed into small balls, boiled with fat or grease and made into cakes, or mixed with powdered saskatoons and eaten with a spoon. Today, it is added to muffins, breads and granola. These large, distinctive plants provide excellent survival food—all parts are edible, and the roots are available year-round (though they become tough and stringy in summer).

MEDICINE: The large, soft leaves provided poultices for burns, and the sticky sap was used as a topical anesthetic and antiseptic on minor wounds, bites and stings. Balsamroot is said to mildly stimulate the immune system, to aid in coughing up phlegm and to stimulate sweating. The roots were boiled to make medicinal teas for treating tuberculosis, rheumatism, headaches, venereal disease and whooping cough and for increasing urine flow, purging the bowels, and facilitating childbirth. Balsamroot poultices were used to heal sores, wounds, blisters and bruises. The sap has anti-bacterial and anti-fungal properties, and the roots have been used to treat athlete's foot and other skin infections. Balsamroot smoke or steam was used to cure headaches and to disinfect the rooms of sick people.

OTHER USES: Balsamroot was sometimes burned as incense in ceremonies.

DESCRIPTION: Clumped, velvety-gray, perennial herb, 8–28" (20–70 cm) tall, from stout, woody, aromatic taproots. Leaves mainly basal, arrowhead-shaped, 8–12" (20–30 cm) long, 2–6" (5–15 cm) wide, long-stalked. Flowerheads yellow, 2–4$^{1}/_{2}$" (5–11 cm) across, with about 12–22 bright yellow ray florets around a deep yellow button of disc florets, borne singly, in April to July. Fruits hairless, 3–4-sided seed-like achenes $^{1}/_{4}$" (7–8 mm) long. Arrow-leaved balsamroot grows on dry, often stony slopes in foothills and montane zones from BC and Alberta to Colorado.

Coltsfoot

Petasites spp.

Arrow-leaved coltsfoot

FOOD: These plants have a mild, slightly salty flavor. Young flowering stems can be eaten as a vegetable, either roasted, boiled or stir-fried. The leaves can be cooked like spinach, but they become fuzzy and rather felt-like with age. The salt-rich ash of coltsfoot leaves was widely used as a salt substitute. The leaves were either rolled into tight balls and dried before burning, to increase the consistency of the ash, or they were fired while encased in balls of clay to contain the ash.

MEDICINE: Coltsfoot is said to be soothing and calming, because it contains mucilage and compounds with antispasmodic and sedative effects. Dried coltsfoot has been used for 100s of years to make medicinal teas to relieve coughing and pain between the ribs in chest colds, whooping cough, asthma and viral pneumonia. It has also been smoked to relieve chronic coughs. These plants are reported to contain antihistamines plus compounds that impede the nerve impulses triggering coughs. Various coltsfoot extracts have been used to lessen spasms and cramps in the stomach, gall bladder and colon. The leaves were used in poultices and salves for relieving insect bites, inflammation, swellings, burns, sores and skin diseases, and they were also boiled to make a decoction that was applied as a treatment for arthritis. Coltsfoot roots were chewed or made into medicinal teas for treating chest ailments (tuberculosis, asthma), rheumatism, sore throats and stomach ulcers. Crushed roots are said to reduce inflammation, to sedate nerves and to inhibit bacterial growth, and they have been used in poultices on sprains, bruises, scrapes and cuts. It is recommended that these poultices be kept on the injury for at least 30 minutes, so that their compounds can be absorbed into the skin.

OTHER USES: Dried leaves have been used in a variety of smoking mixtures.

WARNING:
Coltsfoot contains alkaloids that may be harmful if eaten in large quantities. Pregnant women should not eat these plants. Strong doses may cause miscarriage. Never collect plants from areas where the water may be polluted.

DESCRIPTION: Perennial herbs with 4–20" (10–50 cm) tall flowering stems from slender, creeping rootstocks. Leaves basal, triangular to kidney-shaped, mostly 4–8" (10–20 cm) wide, long-stalked, white-wooly beneath. Flowerheads white to pinkish, about ³/₈" (1 cm) across, with few ray florets and many disc florets, borne in elongating clusters, in April to July (before the leaves appear). Fruits slender, 5–10-ribbed seed-like achenes, with a tuft of white, hair-like bristles.

Arrow-leaved coltsfoot (*P. sagittatus*) has triangular, shallowly toothed leaves. It grows in moist, usually open plains, foothill and montane sites from Alaska to Colorado.

Sweet coltsfoot (*P. frigidus*) has 2 common varieties that are distinguished by leaf shape. **Arctic coltsfoot** (var. *nivalis*) has heart-shaped leaves with coarse teeth and/or shallow lobes, whereas **palmate-leaved coltsfoot** (var. *palmatus*) has rounder leaves that are deeply cut into several finger-like (palmate) lobes. Both varieties are common in the Canadian Rockies, and hybrids are also widespread.

Palmate-leaved coltsfoot

Sweet coltsfoot

Goldenrods

Solidago spp.

Missouri goldenrod

FOOD: Goldenrod plants have been cooked like spinach or added to soups, stews and casseroles. Flavor and texture varies with species, age and habitat. Dried goldenrod leaves or flowers have been used to make pleasant teas (Mormon tea, Blue Mountain tea), usually sweetened with honey. The flowers are edible and provide an attractive garnish for salads. The seeds can be gathered for survival food and used to thicken stews and gravies.

MEDICINE: Goldenrod tea has been used for many years to relieve intestinal gas and cramps, colic and weakness of the bowels and bladder. It is said to reduce the production of mucous in the bronchi, and animal and test tube studies have shown that goldenrod extracts increase the production of urine. Usually goldenrod has been recommended as a cold and flu remedy and as a kidney tonic. Canada goldenrod contains the compound quercetin, which has been used to treat hemorrhagic nephritis (inflammation and bleeding of the kidneys). The tea was also used as a wash for treating rheumatism, neuralgia (pain along a nerve) and headaches, or it was mixed with grease and applied as a salve to sore throats. The flowers were also chewed to relieve sore throats. Powdered leaves and flowers were sprinkled on wounds to stop bleeding, and boiled plants provided antiseptic poultices and lotions. The roots were used for treating burns and relieving toothaches and were also made into medicinal teas for treating colds, kidney stones and painful ulcers. Compounds in goldenrod are believed to stimulate the immune system, so these plants might be used to strengthen allergy defenses at the beginning of hayfever season (like a vaccine). Some herbalists recommend goldenrod tea for relieving exhaustion and fatigue, while others report that it has hypoglycemic activity and is of great benefit to diabetics. During the Crusades, goldenrod was called 'woundwort,' because of its ability to stop bleeding. The genus name, *Solidago*, was taken from the Latin *solidus*, 'whole,' and *ago*, 'to make' (to make whole or cure).

WARNING:
Some people are allergic to goldenrod, so these plants should be used with caution, especially if you are allergic to other plants in the Aster family. People with urinary tract disorders should consult a doctor before using goldenrod. If you are retaining fluids because of a heart or kidney disorder, you should not use this plant. Many people point to goldenrod as a cause of hayfever, but the pollen of these showy flowers is too heavy to be carried by the wind and must be transported by insects. Usually, people are reacting to less conspicuous plants, such as ragweed (*Ambrosia* spp.), that grow in the same habitats.

OTHER USES: The sap of Missouri goldenrod is rich in latex, and attempts have been made to breed cultivars that could be used as a source of rubber. Goldenrod flowers produce a yellow dye with alum as a mordant and a gold dye with chrome as a mordant. Goldenrod is the state flower of Kentucky and Nebraska.

DESCRIPTION: Erect, perennial herbs with alternate, simple leaves. Flowerheads yellow, small (less than $^3/_8$" [1 cm] across), with few (about 10–17) ray florets around a small cluster of disc florets, forming branched clusters, in July to October. Fruits nerved seed-like achenes tipped with a tuft of white, hair-like bristles. The best-known goldenrods are robust, 1–4' (30–120 cm) tall plants with creeping rootstocks.

Of these, **Canada goldenrod** (*S. canadensis*) has finely hairy upper stems and lance-shaped leaves, **giant goldenrod** (*S. gigantea*) has hairless, bluish-green stems, $1^1/_2$–$6^1/_2$' (50–200 cm) tall, and **Missouri goldenrod** (*S. missouriensis*) has hairless, green stems, 8–20" (20–50 cm) tall. All 3 of these species grow on open plains, foothill and montane sites from BC and Alberta to New Mexico.

Northern goldenrod (*S. multiradiata*) has smaller, elongated flower clusters on clumped, 4–20" (10–50 cm) tall, less leafy stems, and its well-developed basal leaves are hairless, except for a distinctive fringe of hairs on their stalks. It grows on open, foothill to alpine slopes from Alaska to New Mexico.

Canada goldenrod (both photos)

185

Curly-cup gumweed

Grindelia squarrosa

FOOD: Some tribes chewed on gumweed leaves or used them fresh or dried to make an aromatic tea.

MEDICINE: Native peoples and early Jesuit missionaries recognized the medicinal value of this plant. Tea made from the resinous flowerheads was taken to relieve indigestion, colic and stomachaches. The leaf tea was used to treat throat and lung problems such as coughs, bronchitis and asthma. It was also widely used as a wash to relieve itching and to heal minor cuts and abrasions, pimples and skin irritations (especially rashes caused by poison-ivy, p. 237). A leaf poultice or the gum from pounded flowerheads was applied to poison-ivy inflammations and minor skin problems. Extracts from the dried flowerheads and leaves are said to have sedative, antispasmodic and expectorant qualities. They were once recommended as a treatment for a wide range of ailments, including headaches, malaria, cancers of the spleen and stomach, gonorrhea, pneumonia, rheumatism, smallpox and tuberculosis. More recently, they have been used in medications for treating asthma and bronchitis, and some herbalists recommend them for treating bladder inflammations caused by fungi or food.

OTHER USES: Plant tops and leaves were used to make washes for healing saddle sores on horses.

DESCRIPTION: Aromatic, biennial or short-lived, perennial herb with branched, 8–24" (20–60 cm) tall stems from taproots. Leaves oblong, hairless, more-or-less lobed to regularly toothed, often clasping the stem and dotted with resinous glands. Flowerheads yellow, ³/₄–1¹/₄" (2–3 cm) across, with 25–40 ray florets around a dense cluster of disc florets and above overlapping rows of backward-curling, sticky involucral bracts, borne in flat-topped clusters, in July to September. Fruits 4–5-ribbed seed-like achenes. Curly-cup gumweed grows on dry, open sites in plains and foothills zones from BC and Alberta to New Mexico.

WARNING:
Large doses of gumweed may cause kidney damage.

Wild sarsaparilla

Aralia nudicaulis

FOOD: These roots were generally considered emergency food only, but some hunters and warriors are said to have subsisted on them during long trips. The fragrant plants have a warm, aromatic, sweetish taste that is most intense in the roots and berries. The roots were used to make tea, root beer and mead. The berries were used to flavor beer and to make wine (similar to elderberry wine), and a beverage tea was sometimes made from the seeds. The berries are generally considered inedible. Some sources report that they have been used to make jelly, but it is not recommended (see Warning). Young shoots were sometimes cooked as a potherb.

MEDICINE: The roots (and occasionally the leaves) were pulverized by pounding or chewing and were used in poultices to soothe and heal wounds, burns, sores, boils and other skin problems and to relieve swelling and rheumatism. Mashed roots were also stuffed into noses to stop bleeding and into ears to stop aching. The roots and berries were boiled to make medicinal teas and syrups or soaked in alcohol to make tinctures. These medicines were used to treat many different problems, ranging from stomachaches to rheumatism and syphilis. The pleasant-tasting root tea was valued as a blood-purifier, tonic and stimulant and as a medicine for stimulating

sweating. It was widely used for treating lethargy, general weakness, stomachaches, fevers and coughs. Wild sarsaparilla was widely used in patent medicines in the late 1800s.

OTHER USES: Wild sarsaparilla roots were boiled with sweet flag (*Acorus calamus*) roots to make a solution for soaking nets to be used for fishing at night. This practice was said to increase the catch.

WARNING:
Some people have reported being very sick after eating wild sarsaparilla berries.

DESCRIPTION: Perennial herb with colonies of single, broad, long-stalked leaves from spreading rootstocks. Leaf blades horizontal, twice divided into 3–5 oval parts 1¼–4¾" (3–12 cm) long. Flowers greenish-white, ¼" (5–6 mm) long, forming 2–7 (usually 3) round, ¾–2" (2–5 cm) wide clusters, in May to June, usually hidden under the leaf. Fruits dark purple berries, ¼–⅝" (6–8 mm) long. Wild sarsaparilla grows on moist, shaded sites in foothills to montane zones from BC and Alberta to Colorado.

Sitka valerian

Valerians

Valeriana spp.

FOOD: Edible valerian was a staple food for some tribes. The roots were usually buried in fire pits and cooked for about 2 days. They were also boiled for long periods in soups and stews. Cooked roots were sometimes dried and ground into flour. Some people find the peculiar taste and smell of these roots offensive, but others enjoy the unusual flavor. Prolonged cooking removes most of the taste and odor. The seeds were sometimes eaten raw, but were said to be better parched.

MEDICINE: Valerian root has been used for centuries as a tranquilizer with fewer side effects than synthetic sedatives. Pharmacological studies have shown valerian extracts to be sedative and antispasmodic when a person is agitated, but a stimulant in response to fatigue. These plants are also said to have anti-bacterial, anti-diuretic and liver-protective properties. All species can be used in much the same way, but some are stronger than others. Sitka valerian and marsh valerian are much stronger than the European species, common valerian (*V. officinalis*), and edible valerian is about half as strong. Valerian root has generally been recommended for relieving stress-induced anxiety, insomnia and muscle tension. It has also been used to treat menstrual problems, epileptic seizures, Saint Vitus's dance and various nervous disorders. Common valerian was widely used during WWII to calm nervousness and hysteria. Native peoples rubbed the chewed or pounded roots onto the head and temples to relieve headaches and applied them to the bodies of people suffering from seizures. The roots were also used in poultices to cure earaches and to heal cuts and wounds. Powdered roots were added to smoking mixtures to relieve cold symptoms.

OTHER USES: Valerian was sometimes used as a perfume in Europe in the 1500s and in the Orient. Cats are said to be very fond of the smell and will dig up the plants and roll on them. Hunters washed their bodies with valerian, believing that this smell would make the deer tamer and easier to approach. Some tribes used dried valerian leaves and roots to flavor tobacco (*Nicotiana tabacum*).

DESCRIPTION: Strong-smelling, perennial herbs with 4-sided stems and opposite leaves. Flowers pinkish to white, tubular, 5-lobed, forming dense (usually), branched clusters. Fruits ribbed seed-like achenes $^1/_8$–$^1/_4$" (2.5–6 mm) long, tipped with many feathery hairs.

Sitka valerian (*V. sitchensis*) has short-lobed, about $^1/_4$" (4–8 mm) long flowers and 3–5-lobed leaves. It grows on moist foothill to alpine slopes from the Yukon and NWT to Idaho and Montana.

Marsh valerian (*V. dioica*) has $^1/_8$" (1–3 mm) long flowers, undivided basal leaves and 9–15-lobed stem leaves. It grows in moist meadows from the Yukon and NWT to Wyoming.

Edible valerian (*V. edulis*) has slender, lance-shaped basal leaves, elongated flower clusters and large, stout taproots. It grows on moist to slightly dry, montane and subalpine slopes from southern BC to New Mexico.

WARNING:
Large doses can cause vomiting, stupor and dizziness. The roots of edible valerian were said to be poisonous unless they had been properly prepared (i.e., cooked for 2 days). Valerian tea should never be boiled. Constant use over a long period of time can cause emotional instability and depression. Many people find the strong odor of these plants (dried, frozen or bruised) nauseating.

Sitka valerian (all images)

Water-parsnip

Sium suave

ALSO CALLED: hemlock water-parsnip.

FOOD: Many tribes gathered water-parsnip roots in spring and early summer and ate them raw, roasted or fried. The crisp, fresh roots were said to have a carrot-like flavor. Because the roots were often collected in spring, before the leaves had grown, plants had to be identified using their roots and the remnants of stems from the previous year. Tender young shoots were sometimes eaten, but mature plants were not considered edible.

MEDICINE: The roots were used to treat stomach problems.

OTHER USES: Children sometimes used the hollow stems to make whistles.

DESCRIPTION: Hairless, perennial herb with hollow, ridged stems 1½–3½' (50–100 cm) tall. Leaves pinnately divided into 5–17 slender, 2–4" (5–10 cm) long leaflets edged with sharp teeth. Flowers white, tiny, forming twice-divided, flat-topped clusters (umbels) 2–7" (5–18 cm) across, in June to August. Fruits pairs of flattened, ribbed, seed-like schizocarps ⅛" (2–3 mm) long. Water-parsnip grows in wet sites in plains, foothills and montane zones from BC and Alberta to New Mexico.

WARNING:
Although the stems and roots are edible, the flowerheads may be poisonous. These plants have been confused with the extremely poisonous water-hemlocks (p. 241), with fatal results. Water-hemlocks are recognized by their twice- (rather than once-) divided leaves, their yellowish, strong-smelling (rather than white, sweet-smelling) roots and their smooth (rather than ribbed) stems. If there is any question about the identity of a plant, consider it poisonous.

Cow-parsnip

Heracleum maximum

ALSO CALLED: H. lanatum
• H. sphondylium.

FOOD: Cow-parsnip was widely used by native peoples. Young stems were gathered before flowering and peeled to remove the strong-smelling outer skin, and the inner stem was eaten raw, steamed or boiled. Tasting like celery and with a texture like rhubarb, it has replaced celery in many recipes and been served as a hot vegetable or used as a tasty snack, raw and dipped in sugar. Unpeeled stems were sometimes roasted in hot coals. More recently, cow-parsnip has been frozen, canned or dried. The roots were used as a cooked vegetable, like parsnips. They are said to taste like rutabagas.

MEDICINE: Cooked roots were eaten to relieve gas, colic and cramps or were mashed as a poultice for drawing boils. Fresh roots have been pounded to a pulp and used as poultices on sores; steeped in hot water to make tea for treating sore throats, coughs, headaches, colds and flu; and boiled to make tea for treating epilepsy, asthma and nervous disorders. Some herbalists have recommended alcoholic extracts of the fresh roots to stimulate nerve growth after injury. Tinctures of the fruits were applied to gum abscesses and toothaches. These plants contain psoralen, a compound that is being studied for the use in treating psoriasis, leukemia and AIDS.

WARNING:
Cow-parsnip could be confused with its poisonous relatives, the water-hemlocks (p. 241). When collecting cow-parsnip, cover your skin. People with sensitive skin may develop dark blotches, rashes and even blisters when contact with this plant is accompanied by exposure to sunlight. These marks can remain for weeks or even months. Furano-coumarins in cow-parsnip have been shown to cause cancer and trigger dangerous cell changes in some animals.

OTHER USES: Toy flutes and whistles can be made from the dry, hollow stems, but they may irritate the lips.

DESCRIPTION: Robust, pungent, perennial herb, $3^1/_2$–$8^1/_4$' (1–2.5 m) tall, with hollow stems. Leaves divided into 3 large (4–12" [10–30 cm] wide), somewhat maple leaf–shaped leaflets and with inflated, clasping stalk bases. Flowers white, forming large (4–12" [10–30 cm] wide), twice-divided, flat-topped clusters, in late May to early July. Fruits flattened, seed-like schizocarps $^1/_4$–$^1/_2$" (7–12 mm) long, with few ribs and 2 broad wings. Cow-parsnip grows on moist sites in plains, foothills, montane and subalpine zones from Alaska to New Mexico.

Angelicas

Angelica spp.

FOOD: Angelica leaves smell rather like parsley and have a strong but pleasant taste, similar to that of lovage (*Levisticum officinale*). The seeds taste like a cross between celery (*Apium graveolens*) and cardamom (*Elettaria cardamomum*). The leaves have been used as a spice or garnish in soups and stews and occasionally as a vegetable. Angelica has also been used to flavor gin and liqueurs. Stems of North American angelicas can be candied like those of European species, but the roots are too tough to be used this way.

MEDICINE: Teas and extracts made from the roots and seeds of angelica have been used for centuries to aid digestion and relieve nausea and cramps. The seeds were said to be most effective for these purposes, whereas the roots had stronger antispasmodic effects. Angelica-root tea has been taken to relieve dry, spastic asthma, intestinal cramps, gas, menopausal discomforts, menstrual cramps and constipation and bloating associated with premenstrual syndrome. It is said to have the advantage of relieving spasms without sedating or stimulating the patient. The Chinese herbal medicine, Dong Quai, which is made from cured plants of *Angelica polymorpha* and/or *Angelica sinensis*, is widely used to treat problems with the digestive tract and the female reproductive system.

White angelica (both photos)

DESCRIPTION: Robust, often ill-smelling, perennial herbs with stout stems from taproots. Leaves alternate, pinnately divided into sharp-toothed leaflets about 1¼–4" (3–10 cm) long and with inflated, clasping stalk bases. Flowers usually white, tiny, borne in large, flat-topped clusters, in June to August. Fruits flattened, winged and ribbed seed-like schizocarps ⅛–¼" (3–6 mm) long. Both common species have hairless fruits in clusters without bracts at the base.

White angelica (*Angelica arguta*) has leaves that are twice divided in 3s. It grows in moist sites in montane and subalpine zones from BC and Alberta to Utah and Wyoming.

Pinnate-leaved angelica (*Angelica pinnata*) has leaves that are mostly divided once into narrower, more lance-shaped leaflets. It grows on similar habitats from Montana to New Mexico.

WARNING:
The roots may be toxic when eaten fresh. Although angelica does not particularly stimulate the uterus, it is probably wise not to use it during pregnancy. Angelicas can cause severe sunburn, rash or other skin reactions on some people when contact or ingestion is followed by exposure to sunlight. Chemicals in the plant have been shown to cause cancer and trigger dangerous cell changes in some animals. Accurate identification is essential to avoid confusion with the extremely poisonous water-hemlocks (p. 241).

White angelica (all images)

Canby's lovage

Ligusticum canbyi

ALSO CALLED: osha.

FOOD: These fragrant plants taste much like they smell. All parts are edible, but the plants have usually been used in small amounts as a flavoring because they tend to be rather overwhelming alone.

MEDICINE: This popular medicinal herb is still widely gathered and traded by some native peoples. The fragrant dried roots were chewed to soothe a wide range of ailments, including sore throats and colds, toothache, headache, stomachache, fever and heart problems. They have also been used to make medicinal teas for treating colds, coughs, flu and sore throats and for use as ear drops. Some present-day herbalists recommend this plant for treating deep, raspy, unproductive coughs caused by viral infections. Roots were mashed or cut into fine shavings and used in poultices on abscessed ears. Some said that a piece of root held over a cut would stop bleeding immediately. To calm seizures, lovage roots were chewed and rubbed on the body and were also smoked in cigarettes. Smoking or chewing these roots has been reported to have a relaxing effect.

OTHER USES: Some tribes used lovage roots and snowbrush leaves (p. 81) to make a hair rinse. Singers chewed lovage roots to give their voices strength and endurance. Root shavings were sprinkled on coals as incense or were used to give a pleasant menthol taste to smoking mixtures of tobacco (*Nicotiana tabacum*) or common bearberry (p. 91). The roots were also used like chewing tobacco—held in the mouth as a snoose. Washing lovage roots was believed to bring rainstorms.

DESCRIPTION: Erect, perennial herb with slender, $3^1/_2$–4' (50–120 cm) tall stems, from a crown of stringy fibers on top of a strong taproot. Leaves fern-like, twice divided in 3s. Flowers white, tiny, forming twice-divided, flat-topped clusters (umbels) about 2–4" (5–10 cm) across. Fruits winged, seed-like schizocarps $^1/_8$" (4–5 mm) long. Canby's lovage grows on moist sites in open foothills and montane zones from southern BC to Idaho and Montana.

WARNING:
Canby's lovage could be confused with some of its extremely poisonous relatives, including water-hemlock (p. 241) and poison hemlock (p. 242). Never eat any plant in the Carrot family, unless you are positive of its identity.

Gairdner's yampah

Perideridia gairdneri

ALSO CALLED: Indian carrot.

FOOD: Yampah was an important food for many native peoples and mountain men. Some people considered these roots the best-tasting wild roots in the mountains, with a sweet, nutty flavor, devoid of bitterness. Some people have likened them to carrots and parsnips, while others say they have a sweet, licorice-like flavor. Yampah roots could be eaten raw, but usually they were boiled, steamed or roasted. They were traditionally collected when the plants were in flower, in spring or early summer. Several tribes dried them for winter use, either whole or boiled, mashed and formed into cakes. Yampah was also stored in earth pits, lined with pine needles (pp. 36–38) or cottonwood bark, where it was safe from frost and rodents. The dried roots were soaked and boiled alone or with other foods, such as with saskatoon (p. 69) and black hair lichen (p. 232). They were also ground into meal and mixed with hot soup or water to make mush or, more recently, mixed with flour to make a thick pudding. Dried yampah root mixed with powdered deer meat was considered a special treat. The seeds have been used as seasoning and were also parched and added to cooked cereals.

MEDICINE: The root tea was used as a laxative, to stimulate urination, to aid in clearing mucous from the lungs and bronchia, and to soothe sore throats.

OTHER USES: The roots were eaten by travelers, hunters and runners and fed to horses to increase energy and improve endurance. This effect was probably owing to the presence of rapidly assimilated sugars and starches.

DESCRIPTION: Slender, 1¼–4' (40–120 cm) tall perennial with a caraway-like fragrance and 2–3 fleshy, tuberous roots. Leaves usually once pinnately divided into slender segments 1–6" (2.5–15 cm) long, but often withered by flowering time. Flowers white, tiny, forming twice-divided, flat-topped clusters (umbels) 1–2¾" (2.5–7 cm) across, in July to August. Fruits brown, rounded, ribbed, seed-like schizocarps about 1/16" (2 mm) long. Yampah grows on open or wooded slopes in plains, foothills and montane zones from southern BC and Alberta to New Mexico.

WARNING:
Never eat any plant in the Carrot family unless you are sure of its identity. There are many poisonous species in this family, and experimentation is like playing herbal roulette.

Fern-leaved desert-parsley

Desert-parsleys
Lomatium spp.

FOOD: The leaves have a strong parsley-like flavor and were often used to flavor meat, stews and salads. The starchy, fibrous roots are edible, but some desert-parsleys may have a strong, resinous, balsam-like flavor. Biscuitroot, with its mild, rice-like flavor, was one of the most popular species. Large roots were sometimes roasted or boiled, but usually they were dried and ground into meal or flour for making mush, bread, cakes and biscuits. Large, flat cakes of biscuitroot meal were carried on long journeys for food. Dried roots and meal were important trade items among some tribes.

MEDICINE: The roots of fern-leaved desert-parsley were used to make a medicinal tea that was considered a general panacea by many native peoples. Most commonly, it was used for treating coughs, colds, sore throats, hayfever, bronchitis, flu, pneumonia and tuberculosis. Some herbalists still use fern-leaved desert-parsley to treat bacterial and viral respiratory infections and to help bring phlegm up from the lungs and bronchia. Extracts from this species have proved effective in combating a wide range of bacteria in laboratory studies.

OTHER USES: Long-distance runners sometimes chewed on desert-parsley seeds to prevent side-aches.

DESCRIPTION: Perennial herbs with a parsley-like fragrance and leafy, branched crowns on long, often tuber-like taproots. Leaves usually divided 2–3 times into small or linear segments. Flowers tiny, borne in open, flat-topped clusters of smaller, round clusters, in April to July. Fruits flattened, winged, seed-like schizocarps about $^1/_4$–$^5/_8$" (5–15 mm) long.

Nine-leaved desert-parsley (*L. triternatum*) has finely hairy, $^1/_2$–4" (1–10 cm) long, linear leaf segments, plus a whorl of bracts at the base of its yellow flower clusters. It grows on open foothill and montane slopes from southern BC and Alberta to Colorado.

Most other species have small, short leaf segments.

Biscuitroot (*L. cous*) is a small plant (usually under 10" [25 cm] tall), with yellow flowers above whorls of broadly oval bracts. It grows on dry, open foothill and montane slopes from Idaho and Montana to Wyoming.

Fern-leaved desert-parsley (*L. dissectum*) has yellow or purplish flowers, slender bracts and corky-winged fruits. It grows on dry, open foothill and montane slopes from southern BC and Alberta to Colorado.

WARNING:
All *Lomatium* species are edible, but as with all members of the Carrot family, these plants can be confused with poisonous relatives. They should be used only when they are positively identified.

Nine-leaved desert-parsley (top)
Fern-leaved desert-parsley (above)

Snakeroot

Sanicula marilandica

MEDICINE: Snakeroot has been reported to have sedative properties, similar to those of valerian (pp. 188–89), for soothing nerves and relieving pain. It was used by native peoples for treating lung, kidney and menstrual problems, rheumatism, syphilis and various skin conditions. It was also said to relieve pain and fever. Herbalists have used snakeroot to cleanse and heal the system, believing that it would stop bleeding, reduce tumors, and heal wounds by seeking out the problem area and focusing its effects there. The root tea was used as a gargle for relieving sore throats, and as a wash for treating skin problems. Its high-tannin content makes it very astringent. The roots were traditionally used as poultices on snakebites, hence the common name 'snakeroot.'

DESCRIPTION: Perennial herb, $1^1/_4$–4' (40–120 cm) tall, from clustered, fibrous roots. Leaves mainly basal, with long-stalked blades $2^1/_2$–6" (6–15 cm) wide, divided into 5–7 sharply toothed, finger-like leaflets. Flowers greenish-white, tiny, forming flat-topped clusters (umbels) of several $^3/_8$" (1 cm) wide, 15–25-flowered heads, in June. Fruits oval, seed-like schizocarps $^1/_8$–$^1/_4$" (4–6 mm) long and covered with hooked bristles. Snakeroot grows on moist, rich, wooded sites in plains and foothills zones from BC and Alberta to New Mexico.

WARNING:
These roots contain irritating resins and volatile oils. Native peoples believed that snakeroot was a powerful medicine, because when its roots were chewed it could cause blistering on the lining of the mouth.

Common plantain

Plantago major

FOOD: It is easy to pull this common weed from the garden, without realizing that it is edible and probably more nutritious than most of the greens we tend. Young leaves have been eaten raw in salads and sandwiches, but they soon become tough and stringy. Cooking improves palatability and makes it possible to remove some of the tougher fibers. Fine chopping may also make older leaves easier to eat. The flavor has been likened to that of Swiss chard (*Beta vulgaris*). Plantain seeds can be dried and ground into meal or flour for use in bread or pancakes.

MEDICINE: Plantains are rich in vitamins A, C and K. The leaves and leaf juice have been widely used in poultices and lotions for treating insect bites and stings, snake bites, sunburn, poison-ivy rashes, sore nipples, blisters, burns and cuts. Plantain leaves have also been heated and applied to swollen joints, sprains, strained muscles and sore feet. In Latin America, common plantain is a prominent folk remedy for treating cancer. Plantain tea has been used for centuries to treat sore throats, laryngitis, coughs, bronchitis, tuberculosis and mouth sores. These plants are said to have anti-inflammatory effects. Also, they contain flavonoids (which are anti-bacterial), allantoin (a soothing compound that promotes healing of injured skin cells) and tannin (whose astringency helps to draw tissues together and stop bleeding). Preliminary studies indicate that plantains may reduce blood pressure, and their seeds have been shown to reduce blood cholesterol levels. Plantain seeds are rich in mucilage, and they were widely used as a source of natural fiber, with laxative effects. They were also used in medicinal teas for treating diarrhea, dysentery, intestinal worms and bleeding of mucous membranes. The roots were recommended for relieving toothaches and headaches and for healing poor gums.

OTHER USES: Strong plantain tea was sometimes used as a hair rinse for preventing dandruff. The tough veins of mature leaves are amazingly strong and have been used as a source of fiber for making thread, fishing line and even cloth. Plantain seeds, soaked in water, produce a mucilaginous liquid that was used as a wave-set lotion.

DESCRIPTION: Clumped, perennial herb with a basal rosette of oval, 5–7-ribbed leaves $1^5/_8$–7" (4–18 cm) long and abruptly tapered to winged stalks. Flowers greenish, about $1/_{16}$" (2 mm) across, with 4 whitish petals, forming dense, slender, $1^1/_4$–12" (3–30 cm) long spikes, in May to September. Fruits membranous capsules $1/_8$" (2–4 mm) long, opening by a lid-like top to release tiny, dark seeds. This European weed grows on disturbed, cultivated or waste ground in plains, foothills and montane zones from Alaska to New Mexico.

Stinging nettle

Urtica dioica

FOOD: Most of the stinging compounds in nettles are destroyed by cooking or drying, but eating large quantities may still cause a mild burning sensation. Tender, young shoots are delicious, boiled and eaten like spinach or added to soups and stews. Nettle purée and cream-of-nettle soup are vivid green. Nettle cooking water has been flavored with lemon and sugar as a hot drink or with salt, pepper and vinegar as a soup base. Young plants have also been used to make nettle tea, wine or beer. Older plants become fibrous and gritty from an abundance of small crystals. Nettle juice or strong nettle tea produces rennet, which was used to coagulate milk to make junket or cheese. Nettle juice has also been used for making beer. The roots, gathered in autumn to spring, were cooked as a starchy vegetable.

MEDICINE: Nettles are rich in protein, minerals and vitamins A and C. The leaf tea is rich in iron, and it is also said to aid coagulation and the formation of hemoglobin. Studies suggest that nettle depresses the central nervous system, inhibits the effects of adrenaline, increases urine flow and kills bacteria. It has been used to treat a wide range of ailments, including gout, anemia, poor circulation, diarrhea and dysentery. Nettle tea was given to women in labor to 'scare the baby out.' It was also believed to increase milk production in nursing mothers and to reduce bleeding associated with menstruation, bladder infections and hemorrhoids. More recently, nettle has been recommended for reducing obesity and for relieving bronchitis, asthma, hives, hayfever, baldness, kidney stones, urinary tract infections, premenstrual syndrome, benign prostatic hypertrophy (noncancerous prostate enlargement), gout, sciatica and multiple sclerosis. In mouthwashes, it can combat dental plaque and gingivitis. Nettle has been used in Germany in

WARNING:
Always wear gloves and long sleeves when collecting nettles. The swollen base of each tiny, hollow hair contains a droplet of formic acid, and when the hair tip pierces you, the acid is injected into your skin. The acid can cause itching and/or burning for a few minutes to a couple of days. Rubbing nettle stings with the plant's own roots is supposed to relieve the burning, and it may have psychological benefits as well. Dock (p. 210) leaves have also been recommended as an antidote. Nettle should not be used by pregnant women (it has caused uterine contractions in rabbits) or diabetics (it has aggravated the diabetic condition of mice).

the treatment of prostate cancer and in Russia for treating hepatitis and inflammation of the gall bladder. These plants are high in boron, which has been reported to elevate estrogen levels, thereby improving short-term memory and elevating the mood of people suffering from Alzheimer's disease. Urtication (stinging the skin with nettles) has been used around the world for centuries to treat rheumatism, arthritis, paralysis and, more recently, multiple sclerosis. Its effectiveness can vary greatly from one case to the next. Stinging could act as a counter-irritant, creating minor pain that tricks the nervous system into overlooking the deeper pain. The stingers also inject a mixture of chemicals into the skin. Some of these chemicals cause inflammation, which could trigger the body to release more of its own anti-inflammatory chemicals.

OTHER USES: Nettle fibers were used for many years to make fishing nets, rope, paper and cloth. The fibers were considered superior to cotton for making velvet or plush and more durable than linen. The roots can be boiled to make a yellow dye or a rinse for reducing hair loss. Roman soldiers, chilled with cold, rubbed their feet and hands with nettles to bring back the circulation.

DESCRIPTION: Erect, perennial herb, armed with stinging hairs, with 4-sided, $1^1/_2$–10' (50–300 cm) tall stems. Leaves opposite, slender-stalked, narrowly lance-shaped to heart-shaped, $1^5/_8$–6" (4–15 cm) long and coarsely saw-toothed. Flowers greenish (sometimes pinkish), $^1/_{16}$" (1–2 mm) long, with 4 tiny sepals and no petals, borne in hanging clusters from upper leaf axils, in April to September. Fruits lens-shaped seed-like achenes $^1/_{16}$" (1–2 mm) long. Stinging nettle grows on moist, rich, often disturbed ground in plains, foothills and montane zones from the southern Yukon and NWT to New Mexico.

Strawberry-blite

Chenopodium capitatum

FOOD: Young plants and leaves can be eaten like spinach—raw or cooked, alone or in stews, soups and salads. They are said to be especially good wilted and tossed with vinegar and bacon. If the leaves are harvested gradually, plants continue to produce all summer. The red, fleshy flower clusters make a refreshing trail-side nibble.

MEDICINE: Strawberry-blite is rich in calcium, protein and vitamins A, B_1, B_2, B_6 and C. As medicine, it was used primarily as a dietary aid for treating diseases caused by nutrient deficiencies. Plants were boiled to make medicinal teas for healing ulcers in the mouth and throat and for using in enemas and douches to treat bleeding bowels and vaginal infections. The cooked leaves are a mild laxative.

OTHER USES: Some native groups used the red flower clusters to make ink or to dye porcupine quills, clothes, hides, basket materials, implements and even their own skin. The color is bright red at first, but it eventually darkens to purple or maroon. In the early 17th century, strawberry-blite was introduced to Europe as an ornamental food plant.

DESCRIPTION: Fleshy, yellowish-green, annual herb with mostly branched, 4–20" (10–50 cm) tall stems from slender taproots. Leaves slender-stalked, broadly arrowhead-shaped, $3/4$–4" (2–10 cm) long, toothless or wavy-toothed. Flowers deep red, pulpy, with 3–5 tiny, fleshy sepals and no petals, borne in dense, $1/2$–$5/8$" (1–1.5 cm) wide heads, in June to August, often forming interrupted spikes. Fruits tiny, with 1 black, lens-shaped seed enclosed in the red, fleshy calyx. Strawberry-blite grows on open, often disturbed ground in foothills, montane and subalpine zones from the Yukon and NWT to New Mexico.

WARNING:
Strawberry-blite can be enjoyed in moderation, but, like its relative spinach (*Spinacia oleracea*), it contains oxalates. Oxalate levels can be reduced by parboiling and by serving these greens with calcium-rich foods such as cream sauces. Cooled ashes have been suggested as an emergency source of neutralizing calcium. The pollen can cause allergic reactions such as hayfever and bronchial asthma.

Lamb's-quarters

Chenopodium album

ALSO CALLED: pigweed.

FOOD: Some consider young, tender, mild-flavored lamb's-quarters superior to spinach. It can be eaten raw, steamed or boiled, but, like spinach, it loses bulk when cooked. Most plants produce new shoots all summer, and flowering tops can also be used. Mature plants were threshed and winnowed to collect the fruits, which were lightly ground and winnowed again to collect the seeds. These seeds were eaten whole or ground into a black flour. It was best to cook and mash the hard, dry seeds before trying to grind them. The flour was often mixed with corn or sunflower-seed flour, because it had a bitter taste alone. It has been used in pancakes, muffins and cookies. The seeds can also be cooked as porridge or used to grow sprouts. Flower clusters can be stripped off and added to soups and salads or eaten with milk and sugar—like cold cereal.

MEDICINE: Lamb's-quarters is rich in vitamins A and C and calcium and is also mildly laxative. The leaves were used as poultices on burns, swellings, wounds and inflamed eyes. They were also bruised and applied to the heads of people suffering from headaches, heat stroke and dizziness. Chewed leaves helped to relieve aching teeth. Lamb's-quarters tea was taken to relieve stomach pains, rheumatism and arthritis.

OTHER USES: The crushed roots have been used as a mild substitute for soap.

WARNING:
These plants should be enjoyed in moderation. Like their relative spinach (*Spinacia oleracea*), they contain oxalates. Parboiling reduces oxalate levels, and calcium counteracts oxalates, so cooked greens with calcium-rich foods (e.g., in a cream sauce) would be safest when eating large amounts of this herb. Cooled ashes are suggested as an emergency source of neutralizing calcium. Some people are allergic to the pollen of lamb's-quarters. Others have skin reactions when exposed to sunlight after eating large quantities of it.

DESCRIPTION: Fleshy, grayish-mealy, annual herb with erect, usually branched stems 8–80" (20–200 cm) tall from slender taproots. Leaves alternate, lance-shaped to diamond-shaped, 3/4–4" (2–10 cm) long, wavy-edged or irregularly lobed. Flowers pale gray-green, with 5 tiny sepals and no petals, borne in spikes of small, mealy, round heads, in July to September. Fruits tiny, containing 1 shiny, black, lens-shaped seed in a thin, papery envelope. Native to Eurasia, this introduced weed grows on disturbed, cultivated or waste ground in the plains, foothills and montane zones from BC and Alberta to New Mexico.

Arrowheads
Sagittaria spp.

FOOD: The entire rootstock of this aquatic plant is edible, but usually only the tubers were used. These tubers are said to have an unpleasant taste when raw, but to be similar to chestnuts (*Castanea* spp.) or potatoes (*Solanum tuberosum*) when cooked. Dried tubers were boiled, but fresh ones were usually roasted in hot ashes or pit-steamed. Sometimes they were cooked with venison and maple syrup. When the syrup was boiled down, they became like candied sweet potatoes. The tubers break away from the main roots easily, so women usually gathered them by hanging onto a canoe and rooting them out with their toes. Dislodged tubers floated to the surface and were tossed into the canoe. Muskrat caches sometimes provided a bushel or more of tubers. Wapato was most widely used, because its ³/₄–2" (2.5–5 cm) long tubers are relatively large. Sometimes, families owned wapato sites and camped nearby for weeks, harvesting the crop. Raw, unwashed tubers can keep for several months, but often they were cooked, sliced and dried for storage. The cooked stems are also said to be tasty.

MEDICINE: Roots were steeped in hot water to make medicinal teas for relieving indigestion. Mashed roots were applied to wounds and sores, whereas leaves were used as poultices to stop milk production. The leaf tea was used to relieve rheumatism and to wash babies with fevers.

Arum-leaved arrowhead (top); Wapato (above)

DESCRIPTION: Aquatic, perennial herbs with milky juice and spreading rootstocks bearing fleshy tubers (often several feet from the parent plant). Leaves basal, arrowhead-shaped (submerged leaves grass-like, up to ³/₈" [1 cm] wide). Flowers white, with 3 broad petals, in whorls of 3 (usually), forming elongated clusters, in June to September. Fruits flattened, short-beaked seed-like achenes in round heads.

Wapato or **broad-leaved arrowhead** (*S. latifolia*) is a large species with leaves up to 10" (25 cm) long, stalks to 3' (90 cm) tall and seed heads over ⁵/₈" (15 mm) wide.

Arum-leaved arrowhead (*S. cuneata*) has ¹/₂–4³/₄" (1–12 cm) long leaves, 8–20" (20–50 cm) tall stalks and seed heads less than ⁵/₈" (15 mm) wide.

Both species grow in calm water in plains, foothills and montane zones from BC and Alberta to New Mexico.

WARNING: Some species of *Sagittaria* cause skin reactions. Wapato is much less common now because of livestock and development pressures. Cattle enjoy the leaves, and pigs eagerly root out the tubers.

Broad-leaved water-plantain

Alisma trivale

ALSO CALLED: *A. plantago-aquatica.*

FOOD: The bulbous plant bases are starchy and edible. They can be very strong-flavored and biting when raw and were said to be best dried thoroughly and then cooked as a starchy vegetable.

MEDICINE: Native peoples used the roots to make medicinal teas for treating lung and kidney problems and lame backs. They also applied the roots as poultices to bruises, sores, swellings, ulcers and wounds. In China these plants are used to increase urine flow in the treatment of dysuria (painful urination), edema (water retention), bladder distention, diarrhea, diabetes and many other problems. Studies there have verified the diuretic action of these plants. Laboratory studies have also shown that water-plantain can lower blood pressure, reduce blood glucose levels, and inhibit fat storage in the livers of experimental animals. Some Chinese people believe that these plants stimulate female reproductive organs and promote conception, whereas the seeds promote sterility. Russians believed that broad-leaved water-plantain would cure rabies.

DESCRIPTION: Semi-aquatic, perennial herb with leafless flowering stems 1–3$^{1}/_{2}$' (30–100 cm) tall from fleshy, corm-like bases. Leaves basal, oval to lance-oblong, with rounded or heart-shaped bases, on long, sheathing stalks. Flowers white, about $^{3}/_{8}$" (1 cm) across, with 3 broad petals, short-lived, borne in several whorls that form elongated clusters, in June to September. Fruits flattened, grooved, oval seed-like achenes in single rings. Broad-leaved water-plantain grows in marshes and ponds from BC and Alberta to Colorado.

WARNING:
Most herbalists consider these plants edible, but some sources describe them as poisonous to people or livestock. Grieve (1931) reported that the leaves of European plants were used as a rubefacient (to redden skin) and could even blister skin, so these plants may cause skin reactions in some. Never eat aquatic plants from polluted water.

American bistort

Bistorts

Polygonum spp.

FOOD: Native peoples ate these starchy rootstocks raw, boiled in stews and soups, or steeped in water and then roasted or dried and ground into flour for bread. The rootstocks can have a pleasant taste, similar to that of almonds or water chestnuts, but occasionally they contain large amounts of tannic acid (as much as 20 percent), making them quite bitter. Mild rootstocks have been suggested as a substitute for raisins or nuts in baking. The leaves and shoots are also edible, with a pleasing tart taste, though flavor and texture vary with age, species and habitat. They have been eaten raw or cooked as a potherb, but the mature stems are usually tough. Like their relative, rhubarb, bistorts can also be sweetened with sugar and cooked as a dessert or made into jam. The seeds are also edible. They have been roasted whole or ground into meal or flour for adding to breads or thickening stews. The tiny bulblets produced by alpine bistort have a pleasant nutty flavor and can be eaten raw, as a trail nibble, or cooked in soups and stews.

MEDICINE: Dried, powdered roots or alcohol or water extracts from the roots are very astringent. They have been used as washes, gargles, douches and enemas to stop bleeding, reduce inflammation and combat infection. These preparations have been used to treat cuts, abrasions and other minor skin problems, infected gums and mouth sores, pimples, measles, insect stings and snake bites. They were also taken internally to treat jaundice and intestinal worms. Bistorts are rich in vitamin C, and the shoots and/or roots were eaten to prevent and cure scurvy.

WARNING:
Although no *Polygonum* species are considered poisonous, their sap is often quite acidic, and raw plants can cause intestinal disturbance and diarrhea when eaten in large quantities. Do not confuse American bistort with the widespread wetland plant water smartweed (*P. amphibium*), which has large (3–6" [8–15 cm] long), often floating leaves and showy, rose-pink flower spikes ¹/₂–1¹/₄" (1–3 cm) long and about ⁵/₈" (1.5 cm) thick. As the name suggests, water smartweed rootstocks will leave your tongue stinging.

DESCRIPTION: Erect, perennial herbs with thick rootstocks. Leaves alternate, mainly basal, elliptic to lance-shaped, with a sheath (fused stipules) at the base of each stalk. Flowers small, with 5 oblong sepals and no petals, forming a single, dense, spike-like cluster, in May to September. Fruits smooth, 3-sided seed-like achenes.

Alpine bistort (*P. viviparum*) is 4–10" (10–25 cm) tall, with narrow ($^1/_4$" [5–8 mm] thick), white flower spikes producing dull brownish achenes near the tip and vegetative bulblets near the base.

American bistort (*P. bistortoides*) is 8–28" (20–70 cm) tall, with showy (at least $^3/_8$" [1 cm] thick) spikes that produce shiny, pale brown achenes (no bulblets).

Both species grow on moist, open slopes in montane, subalpine and alpine zones from BC and Alberta to New Mexico. Alpine bistort is also common farther north.

Alpine bistort with bulblets

Water smartweed (see Warning)

Knotweeds

Polygonum spp.

Mountain knotweed

FOOD: Knotweeds belong to the Buckwheat family, and, like buckwheat, they have seeds that can be eaten whole or pounded into meal and added to breads and sauces. The plants have also been cooked and eaten, though flavor and texture vary greatly with age, species and habitat. Always test a little for palatability before gathering these plants.

MEDICINE: Common knotweed juice was used to stop bleeding, especially nosebleeds. Teas made from these plants were applied externally to hemorrhoids to stop bleeding and were taken internally to relieve diarrhea and treat kidney problems. Because of knotweed's high tannin content and astringency, it may be effective in treating these ailments. An extract made from oak bark and common knotweed was used as a substitute for quinine. There is a long history of the use of knotweeds, including common knotweed, for treating various forms of cancer.

OTHER USES: Elizabethans believed that if the sprawling plants of common knotweed were eaten, they would stunt growth.

DESCRIPTION: Slender, annual herbs with sheathing leaf bases (at least when young) and swollen joints. Flowers inconspicuous, white to pinkish, with 5 sepals and no petals, borne in small clusters of 1–4 at leaf axils and in loose, slender spikes.

Common knotweed or **yard knotweed** (*P. aviculare* or *P. arenastrum*) has small, blue-green, oblong leaves, erect fruits and spreading stems that lie on the ground. These plants were introduced from Europe and have spread as weeds across North America.

Mountain knotweed (*P. douglasii*) is a native species with erect, 4–16" (10–40 cm) tall stems, oval lower leaves, lance-shaped upper bracts and downward-pointing fruits. It grows on dry slopes in plains to subalpine zones from Alberta to New Mexico.

> **WARNING:**
> Although none of these species is considered poisonous, their sap is often quite acidic, and raw plants can cause intestinal disturbance and diarrhea when eaten in large quantities. Some plants contain the toxin hydrocyanic acid. People with sensitive skin should avoid knotweed.

Sheep sorrel

Rumex acetosella

FOOD: These vinegary leaves remain fairly tender until the plant flowers. They have been nibbled as a thirst-quenching trail snack or used to add zest to soups, salads and casseroles. They can also be boiled (sometimes requiring 1–2 changes of water, depending on how sour they are) and served as a hot vegetable. Sheep sorrel steeped in hot water and sweetened with honey can make a refreshing tea, and plants have also been simmered, strained and chilled to make a lemonade-like drink. Cucumber pickles, made with brine and large amounts of sheep sorrel, were considered a delicacy in the early 1900s.

MEDICINE: These nutritious plants, rich in vitamins A and C, were used as a spring tonic and a cure for scurvy. Poultices containing the roasted leaves or the plant juice have a long history as a folk-cure for cancer. The leaves were used to make gargles for treating mouth sores and sore throats, and they were sometimes dried, powdered and dusted onto sores. The leaves and roots were also said to lower fevers, stimulate appetite, reduce inflammation and increase urination. The roots were taken to slow menstrual bleeding, and the seeds were used to treat bleeding, diarrhea and dysentery. Some present-day herbalists consider sheep sorrel a blood alterative and liver stimulant, useful for treating skin problems and other metabolic imbalances. The plants contain compounds called anthraquinones, which have laxative, anti-bacterial and anti-fungal properties.

WARNING:
Sheep sorrel (like its relative, spinach [*Spinacia oleracea*]) contains oxalates, which irritate the digestive tract and can be toxic, although a large quantity would have to be eaten to cause poisoning. Acid levels can be reduced by parboiling. Also, calcium helps to neutralize oxalates, so the hazard can be further reduced by serving sorrel with calcium-rich foods (e.g., in a cream sauce). Cooled ashes have been suggested as a possible emergency source of neutralizing calcium. Elderly people, small children and people with kidney problems, gout or rheumatism should not use plants with oxalates. People who are allergic to grass, tree or shrub pollen may also be sensitive to sheep sorrel.

DESCRIPTION: Slender, hairless, perennial herb, about 8–12" (20–30 cm) tall, with long-stalked, arrowhead-shaped leaves about $3/4$–$1^5/8$" (2–4 cm) long. Flowers reddish (sometimes yellowish), either male or female, forming loose, branched clusters, in April to August. Female flowers have 3 broadly oval scales (valves) $1/16$" (1–2 mm) long and 3 scale-like sepals, whereas male flowers have 6 scale-like sepals. Fruits glossy golden-brown, 3-sided seed-like achenes, closely enveloped in 3 valves. This European introduction grows on cultivated or waste ground in plains to subalpine zones from BC and Alberta to New Mexico.

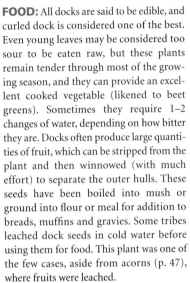

Docks

Rumex spp.

FOOD: All docks are said to be edible, and curled dock is considered one of the best. Even young leaves may be considered too sour to be eaten raw, but these plants remain tender through most of the growing season, and they can provide an excellent cooked vegetable (likened to beet greens). Sometimes they require 1–2 changes of water, depending on how bitter they are. Docks often produce large quantities of fruit, which can be stripped from the plant and then winnowed (with much effort) to separate the outer hulls. These seeds have been boiled into mush or ground into flour or meal for addition to breads, muffins and gravies. Some tribes leached dock seeds in cold water before using them for food. This plant was one of the few cases, aside from acorns (p. 47), where fruits were leached.

MEDICINE: Docks were often used interchangeably as medicines, but curled dock was one of the most widely recognized and used. The leaves are rich in protein, calcium, iron, potassium and vitamins. Curled dock can contain more vitamin C than oranges and more vitamin A (beta-carotene) than carrots. Dried, powdered leaves were sometimes used as dusting powder, added to salves or applied as a paste for healing sores and reducing itching. Dock leaves were rubbed on nettle stings to relieve the burning, They were also used to make gargles for treating mouth sores. Some present-day herbalists consider curled dock a liver stimulant and blood cleanser, useful for treating liver problems, swollen lymph nodes, sore throats, skin sores, warts and rheumatism. The root tea has been used to treat jaundice and post-hepatitis flare-ups. Some people believe that the tea helps the body to eliminate heavy metals such as lead and arsenic. Dock plants contain anthraquinones, which have laxative and anti-bacterial effects and

Western dock

also stop the growth of ringworm and other fungi. Docks have been widely used as mild laxatives and in washes for treating fungal infections.

DESCRIPTION: Robust, often reddish, perennial herbs, $1^1/_2$–5' (50–150 cm) tall from taproots. Leaves alternate, narrowly oblong to lance-shaped. Flowers small, reddish, with 3 net-veined, heart-shaped scales (valves), hanging in dense, branched clusters, in June to August. Fruits smooth seed-like achenes about $1/_{16}$" (2–2.5 mm) long, covered by the flower valves.

Curled dock (*R. crispus*), a European introduction, has wavy, curled leaf edges, wedge-shaped leaf bases and unbranched stems (below the flower cluster), and each valve has a large, oval bump (tubercle).

Willow dock (*R. salicifolius* or *R. triangulivalvis*) has plane or only slightly wavy leaf edges, wedge-shaped leaf bases and branched stems, and each valve has a lance-shaped bump (tubercle).

Western dock

Western dock (*R. aquaticus* or *R. occidentalis*) has wavy to slightly curled leaf edges and notched leaf bases, and its valves lack tubercles.

All 3 species grow on moist, often disturbed ground in plains, foothills and montane zones from BC and Alberta to New Mexico. Willow dock and western dock are also found farther north or at slightly higher elevations.

WARNING:
These plants (like their relative, the beet [*Beta vulgaris*]) contain oxalates, which are toxic in large quantities, but they are safe to use in moderation. Acid levels can be reduced by parboiling, and calcium helps to neutralize oxalates, so the hazard can be further reduced by serving dock with calcium-rich foods such as cream sauces. Cooled ashes have been suggested as an emergency source of neutralizing calcium. Docks produce large amounts of air-borne pollen that can cause allergic reactions, such as hayfever and asthma.

Mountain sorrel
Oxyria digyna

FOOD: The leaves are tender and rather fleshy, with a sharp, almost vinegary flavor. They make a refreshing, thirst-quenching snack on the trail or a tasty addition to salads and sandwiches. They have also been chopped and soaked in water and then sweetened with sugar to make a drink like lemonade. Mountain sorrel plants are sometimes cooked as a potherb or puréed as a soup thickener. They can make delicious cream soups and purées and can add zip to fish, rice and vegetable dishes. Some tribes fermented mountain sorrel to make a type of sauerkraut, which was dried or stored in seal oil for winter use. Others boiled the plants with berries and salmon roe until the mixture thickened. They then poured the mixture about $3/4$" (2 cm) thick into frames where it dried as cakes. These cakes were often an important trade item. The boiled leaves, sweetened with sugar and thickened with flour, are said to resemble stewed rhubarb. These plants will produce succulent leaves year after year, if their rootstocks are left undisturbed in the ground.

MEDICINE: Mountain sorrel is rich in vitamins A, B and C, and it was widely used to prevent and cure scurvy. It was also taken internally or used in poultices and washes to treat diarrhea, sores, ulcers, rashes, itching, ringworm and even cancer. Some recommend the leaves for easing the itch of mosquito bites.

DESCRIPTION: Tufted, hairless, often reddish, perennial herb, about 4–8" (10–20 cm) tall, from fleshy taproots or rootstocks. Leaves mostly basal, long-stalked, kidney-shaped to heart-shaped, $3/8$–$15/8$" (1–4 cm) across. Flowers greenish to reddish, about $1/16$" (1.5 mm) long, with 4 sepals and no petals, hanging on slender stalks, forming $3/4$–6" (2–15 cm) long, branched clusters, in July to September. Fruits reddish, broadly winged, lens-shaped seed-like achenes $1/8$–$1/4$" (3–6 mm) wide. Mountain sorrel grows on moist, open sites in montane, subalpine and alpine zones from Alaska to New Mexico.

WARNING:
These plants should be eaten in moderation. They contain oxalates, which are toxic when consumed in large quantities (see Warning under sheep sorrel [p. 209]).

Bulrushes

Scirpus spp.

FOOD: The juicy, young shoots and lower stalks have been eaten raw as a thirst-quenching snack or cooked as a vegetable. The starchy rootstocks are best collected in autumn or early spring. Growing tips of the rootstocks have been eaten raw, roasted like potatoes or added to soups and stews. Dried rootstocks were crushed to remove their fibers and then ground into flour for making bread, muffins and biscuits. Similarly, fresh rootstocks have been boiled into a gruel and then dried and ground into flour or used wet in pancakes and breads. Young rootstocks were crushed and boiled to make a sweet syrup. Pollen was pressed into cakes and baked, or it was mixed with flour and meal to make bread, mush or pancakes. The seeds can be eaten raw or parched. They were added to stews, breads and mush or ground into meal and used as a flour supplement or thickener. Some tribes gathered the sweet, dried sap or 'honeydew' exuded by the stems and rolled it into balls for storage.

MEDICINE: Plants were boiled to make medicinal teas for treating nervous, fretful, crying children and for washing weak legs.

OTHER USES: The long, flexible stems of these tall plants can be woven into baskets and mats or twisted to make the seats of rush-bottomed chairs. They were often used with or instead of cattail leaves (pp. 214–15).

DESCRIPTION: Slender, pale green, aquatic herbs with round, 1 1/2–10' (50–300 cm) tall stems from scaly, creeping rootstocks. Leaves mostly reduced to bladeless sheaths. Flowers tiny, in finely hairy, oval spikelets, forming a branched, 2–4" (5–10 cm) long cluster from the base of an erect, stem-like bract, in June to September. Fruits 2-sided, oval seed-like achenes about 1/16" (1.5–2.5 mm) long, with 4–6 slightly barbed bristles from the base.

Common great bulrush (both photos)

Common great bulrush or **tule** (*S. tabernaemontani* or *S. validus*) has soft, spongy stems that are easily crushed. Its 1/16–3/32" (2–2.3 mm) long achenes are covered by reddish-brown scales.

WARNING:
Never eat bulrushes from polluted water.

Hard-stemmed bulrush (*S. acutus*) has firmer, less easily crushed stems. Its 1/16–3/32" (2.2–2.5 mm) long achenes are covered by pale grayish-brown scales with reddish-brown lines.

Both species grow in shallow, calm water from BC and Alberta to New Mexico, and common great bulrush is also found in the southern Yukon and NWT.

Cattails

Typha spp.

Common cattail

FOOD: The tender inner parts of young shoots (with the outer leaves removed) have been likened to celery (*Apium graveolens*) or asparagus (*Asparagus officinalis*) and are good raw, steamed or stir-fried. Green flower spikes have been cooked and eaten like corn on the cob. Oil- and protein-rich cattail pollen has been used to thicken sauces and gravies or mixed with flour to make muffins, biscuits, pancakes and cookies. It can be gathered by shaking pollen-laden spikes into a bag and sieving the powder. The starchy white core of the rootstocks was eaten raw, boiled or baked. It was also dried and ground into flour, boiled to make syrup or fermented to make ethyl alcohol. The starch can be collected by peeling and crushing the roots in water, straining out the fibers, and washing the heavy white starch in several changes of water, carefully pouring off the water each time. The seeds were used in breads and porridges. Burning the fluff both separated and parched the seeds. More recently, cattail seed has been suggested as a source of oil, with the by-products providing chicken feed.

MEDICINE: Cattail was once used as a chew for treating coughs, and the rootstocks were recommended for increasing urination and for treating gonorrhea and chronic dysentery. Green flower spikes were eaten to relieve digestive disorders such as diarrhea. Rootstocks were steeped in water or milk to make medicinal teas for treating abdominal cramps, diarrhea and dysentery. Boiled or raw rootstocks were pounded to a jelly-like paste and applied to sores, boils, wounds, burns, scalds and inflammations. Down from the flower spikes was also used, either dry or in salves, to heal burns, scalds and smallpox pustules. The sticky juice was rubbed on gums as a topical anesthetic when extracting teeth.

OTHER USES: Rootstocks and leaves were used for caulking boats and barrels. Down from flower spikes provided bedding, diapers and baby powder for infants. It was also used to stuff mattresses, pillows, sleeping bags and life jackets, and it provided tinder, insulation and soundproofing. Cattail quilts would not let water penetrate, so they were placed

WARNING:
Mature cattails are unmistakable, but be careful not to confuse young shoots with poisonous members of the Iris family, such as western blue flag (p. 241). Plants in salty or stagnant water often taste like a dirty aquarium.

over mattresses and used for babies. The down was mixed with ashes and lime to make a cement that was said to be as hard as marble. The leaves were woven into bed mats, chair seats, baskets and water jugs, and they were also fashioned into toy figures of humans and ducks. Cattail mats, often several layers thick, covered tipis, sweat baths and Sundance lodges. Mats were sometimes hung in homes to bring rain and to protect the families and their animals from lightning. Historically, cattail pollen was widely used in religious ceremonies, but as agriculture developed, it was replaced with corn (*Zea mays*) pollen. It could be used in pyrotechnics, to produce bright flashes of light. Cattail seeds were said to kill mice. Cattail stems have been used to make a type of glue, and the pulp can be made into rayon.

DESCRIPTION: Emergent, perennial herbs with pithy stems from coarse rootstocks. Leaves sword-like, rather stiff and spongy. Flowers tiny, lacking petals and sepals, forming dense, cylindrical spikes, with yellow male flowers at the tip (soon disintegrating) and green to brown female flowers below, in June to July. Fruits tiny seed-like achenes with a tuft of long hairs, borne in brown spikes.

Common cattail (*T. latifolia*) has large plants (often over 5' [1.5 m] tall) with ³/₈–³/₄" (8–20 mm) wide leaves, and the male and female parts of its spikes are contiguous. It grows in calm water in plains, foothills and montane zones from the southern Yukon and NWT to New Mexico.

Narrow-leaved cattail (*T. angustifolia*) is smaller (3¹/₂–5' [1–1.5 m] tall), with narrower (about ¹/₄" [5 mm] wide) leaves, and the male and female parts of its spikes are usually separated by a ³/₈–2" (1–5 cm) gap. It grows in similar habitats in plains and foothills zones from Montana to New Mexico.

Common cattail

215

Grasses

Poaceae

Reed canarygrass

FOOD: Grasses are mankind's most important food plants. Wheat (*Triticum aestivum*), rice (*Oryza sativa*) and corn (*Zea mays*) are grasses. All native grasses have edible, starchy grains, but species with large grains that separate readily from surrounding bracts (chaff) are, understandably, most popular. Traditionally, flower clusters were beaten over containers, roughened to loosen their bracts, and then winnowed to remove the chaff. Sometimes, a light burning removed chaff and hairs and parched the grains. Cleaned grains were cooked as gruel, added to soups and stews as thickeners, or ground into flour or meal for making bread or gruel. Young tender shoots, large starchy rootstocks and thick stem bases have also been eaten raw, pickled, roasted or boiled as a vegetable, or dried and ground into flour. Rootstocks were also roasted until dark brown, ground and used as a substitute for coffee. Punctured stems of young common reed (before flowering) sometimes exude a sweet, pasty substance that hardens into a gum. This substance was pressed into balls and eaten like candy, or dried stalks were ground into a sweet flour that was moistened and formed into balls. These balls were then heated by the fire until they swelled and turned brown and could be eaten like taffy (an early version of the roasted marshmallow).

MEDICINE: Grasses have been used for many years to treat many ailments—ranging from hiccups to cancer. However, they are not commonly employed by present-day herbalists. Some grass pollen is used in the preparation of vaccines or allergens for treating hayfever.

OTHER USES: Dried leaves of large grasses (e.g., reed canarygrass) were softened by rubbing and stuffed into mattresses and pillows or used as insulation in clothing and foot-gear. Large grasses were twisted together to make cord and nets and were used to make thatched roofs and walls, screens, mats, baskets, prayer sticks, light arrow shafts (for children's bows and small-game hunting), weaving rods, pipe stems and even fishing poles. Some large, abundant grasses (e.g., common reed) have been used as a source of cellulose for manufacturing the textile fiber, rayon. The flower clusters of many grasses dry well and keep their color. These flower clusters can add a wide range of shapes, colors and textures to dried flower arrangements. Grasses provide important forage and hay for livestock. Dense roots and spreading rootstocks of many grasses stabilize soil and prevent erosion.

DESCRIPTION: Large, diverse group of herbs with solid, often swollen joints (nodes) on round, hollow stems. Leaves linear, parallel-veined, in 2 vertical rows, sheathing the stem. Sheaths usually split down 1 side. Flowers (florets) tiny, with 2 specialized scales (an outer lemma and inner palea) enclosing the stamens and/or ovary, borne in spikelets of 1 or more florets above a pair of stiff bracts (glumes), forming clusters ranging from small, compact spikes to large, feathery panicles. Fruits starchy grains.

Reed canarygrass

Reed canarygrass (*Phalaris arundinacea*) is a 2–5' (60–150 cm) tall perennial with thick rootstocks and straw-colored, narrow, branched clusters of flattened, $^{1}/_{8}$–$^{1}/_{4}$" (4–6 mm) long spikelets bearing 1 fertile floret. It grows on moist, open sites (often in ditches) in plains, foothills and montane zones from the southern Yukon and NWT to New Mexico.

Common reed (*Phragmites australis* or *Phragmites communis*) is 6$^{1}/_{2}$–10' (2–3 m) tall, with many broad ($^{3}/_{8}$–1$^{5}/_{8}$" [1–4 cm] wide) leaves below a large (4–16" [10–40 cm] long), pale to purplish, plume-like flower cluster. It grows on wet plains and foothill sites from the southern NWT to New Mexico.

WARNING:
Blackened grains are often infected with ergot (*Claviceps* spp.), a poisonous fungus that can cause severe illness or death. Eating small amounts of ergot for long periods can cause the loss of nails in mild cases and the loss of hands or feet plus gangrene of internal organs in extreme cases. Convulsive ergotism, reported in historical accounts from the 1700s, resulted from eating large amounts of ergot over a short time. The pollen of many grasses is highly allergenic.

217

Common sweetgrass

Hierochloe odorata

FOOD: Although sweetgrass grains are edible, they do not appear to have been used as food. These plants are rich in coumarin and should not be eaten (see Warning).

MEDICINE: Sweetgrass smoke was inhaled to relieve colds. The plants were used to make medicinal teas for treating coughs, sore throats, fevers, venereal infections, chafing, windburn and sore eyes. These plants were also taken to alleviate sharp internal pains, to stop vaginal bleeding and to expel afterbirth.

OTHER USES: Common sweetgrass leaves were collected in autumn and used to perfume clothing and to repel insects and other 'bugs.' Some women wove sweetgrass into their hair or soaked the leaves in water to make a sweet-smelling hair rinse. Sweetgrass was also soaked in water with gelatin from boiled horse hooves to make a hair lotion. Sweetgrass is still commonly burned as incense, and its smoke is used to purify people and implements during religious ceremonies and to bring blessings and protection. Interwoven strands of sweetgrass are a symbol of life's growth and renewing powers. Some tribes chewed sweetgrass during religious fasts for greater endurance or mixed it with tobacco (*Nicotiana tabacum*) for ceremonial smoking. Silvery or golden sweetgrass flower clusters make a lovely, sweet-smelling addition to dried bouquets. The long, creeping rootstocks stabilize unstable slopes and roadsides.

DESCRIPTION: Sweet-smelling, vanilla-scented, perennial grass, 1–2' (30–60 cm) tall from slender, creeping rootstocks. Leaves mainly basal, flat, usually about ¹/₈" (3–5 mm) wide. Flowers in shiny, bronze to purplish, pyramid-shaped clusters of flattened, 3-flowered, tulip-shaped spikelets ¹/₈–¹/₄" (4–6 mm) long, produced in May to July. Common sweetgrass grows on open, moist to dry, often disturbed ground in plains to subalpine zones from Alaska to New Mexico.

WARNING:
Sweetgrass contains the sweet-smelling compound coumarin, which delays or prevents blood from clotting.

Western polypody

Polypodium hesperium

ALSO CALLED: common polypody.

FOOD: The sweet, licorice-flavored rootstocks of western polypody contain sucrose and fructose, as well as osladin, a compound that is said to be 300 times sweeter than sugar. Although these tough rootstocks have seldom been used as food, they were occasionally used for flavoring and they were often chewed as an appetizer. Chewing polypody rootstocks before drinking water was said to make the water taste sweet.

MEDICINE: The rootstocks of a closely related species, licorice fern (*P. glycyrrhiza*), were chewed (raw or roasted) to soothe sore throats and relieve cold symptoms. They were also boiled to produce medicinal teas for treating dysentery, for washing chapped hands and dislocated joints, for expelling intestinal worms and for dispelling depression and bad dreams. Because of their sweet, licorice flavor, polypody rootstocks were sometimes used to mask the taste of other bitter medicines. Present-day herbalists have recommended these rootstocks for reducing inflammation caused by allergic reactions.

OTHER USES: Some settlers used polypody rootstocks to flavor tobacco (*Nicotiana tabacum*).

DESCRIPTION: Evergreen fern with $1^5/_8$–6" (4–15 cm) long leaves scattered along creeping, reddish-brown, scaly rootstocks. Leaves (fronds) lance-shaped in outline, pinnately divided into 5–18 opposite or offset pairs of blunt lobes. Leaf stalks $^1/_2$–$4^3/_4$" (1–12 cm) long, smooth, straw-colored, shorter than the blade. Reproducing by spores borne in large, rounded dots (sori) on the underside of leaflets, midway between the edge and the central vein. Western polypody grows in crevices on rocky foothill and montane slopes from BC and Alberta to New Mexico.

> **WARNING:**
> All ferns should be used with caution and never during pregnancy, because the toxicity of most species is not known. Some *Polypodium* species contain methyl salicylate, a compound with aspirin-like effects, so people who are allergic to aspirin or who have blood-clotting disorders should not use these plants.

Broad spiny woodfern

Woodferns

Dryopteris spp.

FOOD: The bitter rootstocks of narrow spiny woodfern and broad spiny woodfern become sweeter when cooked, and some tribes gathered them for food. Cooking involved several hours of boiling in pots, roasting in or over coals, or steaming in pits. Cooked rootstocks were peeled and eaten with grease or fermented salmon roe. Their flavor has been likened to that of sweet potatoes (*Dioscorea alata*). The fleshy rootstocks were usually gathered in later autumn or early spring (before the leaves developed). Those rootstocks edged with round, fleshy, light green fingers of new growth were eaten, but if the fingers were dark inside, the rootstock was not considered edible. Some say that the raw rootstocks of all 3 species are edible, but consumption is not recommended (see Warning).

MEDICINE: Oil from the rootstocks of male fern has been used for hundreds of years to expel tapeworms from humans and animals. The patient was put on a fat-free diet for 2–3 days and then given a dose of male fern, followed by a salty laxative. A single dose was often enough. Compounds in the resins of these ferns paralyze the worms and force them to release their hold on the intestine. The worms were then flushed from the system by the salty laxative. Male fern was also said to cause abortions. Mashed rootstocks were sometimes applied to skin to heal cuts and ulcers. Some native peoples recommended woodfern rootstocks as an aid for losing weight and a cure for poisoning by the red tide.

OTHER USES: Woodfern leaves were soaked in water to make a hair rinse.

DESCRIPTION: Large, clumped ferns, about 1–3' (30–90 cm) tall, from brown-scaly, stout rootstocks covered with old leaf bases. Leaves (fronds) feathery, with dark, brown-scaly stalks. Reproducing by spores in dots (sori) on the veins, each dot partly covered by a thin, horseshoe-shaped to kidney-shaped flap (indusium).

Two common woodferns have leaves that are divided 2–3 times into small leaflets with softly spine-tipped teeth. **Broad spiny woodfern** (*D. expansa*, *D. assimilis* or *D. dilatata*) has triangular to oblong-lance-shaped leaves with an asymmetrical lowermost pair of leaflets. **Narrow spiny woodfern** (*D. carthusiana*, *D. austriaca* or *D. spinulosa* [in part]) has narrower, nearly symmetrical leaves. Both species grow in shady foothill, montane and subalpine sites from the Yukon and NWT to Idaho and Montana.

Male fern (*D. filix-mas*) has twice-divided leaves with blunt teeth. It grows on moist slopes in foothills, montane and subalpine zones from southern BC and Alberta to New Mexico.

WARNING:
Raw rootstocks are bitter, strongly laxative and potentially toxic. Overdoses cause muscular weakness, coma and, most frequently, blindness. Male fern will irritate sensitive skin.

Bracken

Pteridium aquilinum

FOOD: Young coiled leaves (fiddleheads), 4–6" (10–15 cm) tall were collected in spring and rubbed free of hairs. They were then soaked in salt water, to reduce bitterness, and eaten raw (see Warning). Usually, however, fiddleheads were boiled for 30 minutes in 2 changes of water and served as a hot vegetable. Some tribes also dried them for storage. Rootstocks were roasted or pit-steamed. Cooked rootstocks were peeled and then pounded to separate the whitish, edible part from the fibers. Some people simply chewed small chunks until the starch was gone and spat out the fibers. Bracken rootstocks were often eaten with oily foods, because these roots were said to be constipating. Dried rootstocks were ground into flour, then mixed with water to make dough, formed into cakes, and roasted. Warabi starch, extracted from the rootstocks of bracken, has been used in candies. Bracken rootstocks were also used to make ale.

MEDICINE: Bracken rootstocks were used in teas for treating rickets, stomach cramps, diarrhea and worms. Salves (made by boiling the leaves in fat) or boiled, mashed rootstocks were applied to burns, sores and caked breasts (breasts with milk accumulated in the ducts). The leaves were smoked to relieve headaches.

OTHER USES: Bracken tea was sometimes used in rinses for stimulating hair growth. The tannin-rich leaves were used for tanning leather. They were also used to dye wool yellow-green and to dye silk gray. Burning leaves provided mosquito-repellent smudges. These large, abundant leaves resist decay, so they have been used in many ways, including thatching, bedding and packing material for produce. Rootstocks lather in water and have been used as a soap substitute.

WARNING:
Sheep, pigs, cattle and horses have been poisoned by eating large amounts of mature bracken leaves. These leaves contain thiamase, an enzyme that disturbs thiamine (vitamin B_1) metabolism. Cooking destroys this toxin. Bracken also contains known carcinogens and mutagens. It has been reported to cause cancer in grazing animals, and in Japan it is suspected of causing stomach cancer in people.

DESCRIPTION: Coarse fern, 20–80" (0.5–2 m) tall, with single, spreading to horizontal, long-stalked leaves from deep, spreading rootstocks. Leaves triangular in outline, 1–3$^1/_4$' (30–100 cm) long, 2–3 times pinnately divided into firm, round-toothed leaflets, turning a bright rusty color after freezing in autumn. Reproduces by spores borne in a continuous band under down-rolled leaf edges. Bracken grows in a wide range of sites, often on disturbed ground, in foothills and montane zones from BC and Alberta to New Mexico.

Horsetails & Scouring-rushes

Equisetum spp.

FOOD: The rootstocks and small, brown, fertile shoots of common horsetail have been eaten, fresh or boiled, but they should be used only in moderation, if at all. The tough outer fibers are either peeled off or chewed and discarded. Young, green, tightly compacted bottlebrush shoots were cleaned of their sheaths and branches and eaten occasionally, but consumption is not recommended (see Warning).

MEDICINE: Horsetails and scouring-rushes have been used in much the same manner as medicines, though scouring-rushes were often considered stronger. Teas made from these plants were usually taken to treat bladder and kidney problems, water retention and constipation, but they were also used for treating gout, gonorrhea, stomach problems, menstrual irregularities, bronchitis and tuberculosis and were applied as washes to stop bleeding and combat infection. Plants were used in poultices to relieve bladder and prostate pain and to heal cuts, wounds and sores. Also, their ashes were applied to mouth sores. Horsetails and scouring-rushes may have some antibiotic properties, and glycosides in some plants have a weak diuretic action. Some present-day herbalists value these plants for their highly absorbable silica and calcium content. These minerals are said to maintain the health and resilience of cartilage and connective tissues, to promote circulation in the scalp and to strengthen bones. Horsetails and scouring-rushes have been recommended for preventing osteoporosis (increased bone porosity), for strengthening fingernails

WARNING:
Horsetails have caused deaths of horses and cattle through the action of the enzyme thiaminase, which destroys thiamine (vitamin B_1). Cooking destroys thiaminase, and B-vitamin supplements reverse its effects. Some horsetails also contain alkaloids, such as nicotine and palustrine, which are toxic in large quantities. Large amounts of silica may irritate the urinary tract and kidneys. Horsetails readily absorb heavy metals and chemicals from the soil, so plants from contaminated soil should never be used for food or medicine. People with high blood pressure and related problems should not use horsetails.

Common horsetail (top)
Common scouring-rush (above)

and for treating bursitis (inflamed connective tissue around joints), tendinitis (inflamed tendons), baldness, bone fractures, mouth infections, offensive perspiration and various eye, teeth, nail and skin problems. However, there is still controversy over the efficacy of these plants and the potential dangers associated with their use.

OTHER USES: These plants are embedded with abrasive silica crystals, so they can be used as polishers, facial scrubs and disposable pot scourers. Native peoples gathered scouring-rushes to polish pipes, bows and arrows, and European housewives used these abrasive plants to brighten tins, floors and woodenware. Boxes of scouring-rush segments are still sold for shaping clarinet and oboe reeds. Teething babies were given horsetail rootstocks to chew on, and older children used common scouring-rush stems to make whistles (though their elders warned them that blowing the whistles could attract snakes). Tea made

Common horsetail (sterile stems)

from these plants was used as a hair rinse to reduce baldness and kill lice.

DESCRIPTION: Perennial herbs with hollow, jointed, vertically ridged stems from creeping rootstocks. Leaves reduced to small scales, fused into sheaths on the stem. Reproducing by spores borne in cones (strobili) at the stem tips.

Common scouring-rush (*E. hyemale*) has stiff, unbranched, 18–40-ridged, evergreen stems $^1/_8$–$^3/_8$" (4–10 mm) thick and 1–3$^1/_4$' (30–100 cm) tall, with $^1/_8$–$^3/_8$" (3–10 mm) long sheaths that have 2 black bands. It grows on moist, sandy, often disturbed ground in plains, foothills and montane zones from the Yukon and NWT to New Mexico.

Common horsetail or **field horsetail** (*E. arvense*) produces 2 types of plants—brownish, unbranched fertile plants, 4–10" (10–25 cm) tall (in spring) and green, bottlebrush-like, sterile plants, 4–16" (10–40 cm) tall (in summer). Common horsetail grows on a wide range of sites, in plains to alpine zones from Alaska to New Mexico.

Common horsetail (fertile stems)

Northern fir clubmoss

Clubmosses
Lycopodium spp.

FOOD: These plants have been used occasionally for food, but they are described as rather unappetizing and they can be toxic (see Warning).

MEDICINE: Clubmoss spores have been used as dusting powder on wounds and surgical incisions and as treatment for various skin problems, including eczema, herpes, rashes and chafed skin. They have been used to untangle and delouse matted hair, to prevent pills or suppositories from sticking together, and as a powder on condoms and rubber gloves. Animal and test tube studies have shown that clubmosses contain alkaloids that can increase urine flow, relieve spasms, stimulate bowel movements, vomiting and estrogenic activity (including uterine contractions), reduce fever, pain and inflammation, and combat bacteria and fungi. Running clubmoss plants were used to treat several ailments, including kidney problems (especially those associated with high uric acid concentrations), postpartum pain, fever, weakness, rheumatism and even rabies. The spores were used to stop bleeding and to treat diarrhea, dysentery, rheumatism and indigestion. They were also believed to be an aphrodisiac. Fir clubmosses were recommended for treating a variety of complaints, including uterine problems, knee problems, swollen thighs and water retention. More recently, they have been found to contain the alkaloid huperzine, which has been shown to increase the efficiency of learning and memory in animals by inhibiting the breakdown of acetylcholine, an essential neurotransmitter. These plants are now being studied for use in the treatment of myasthenia gravis (a disease in which muscles become increasingly weak and easily tired) and Alzheimer's disease.

OTHER USES: Clubmoss spores are rich in oil and are highly flammable. At one time, they were used by photographers as flash powder. Theater performers used them to produce special effects such as lightning and explosions. The fine yellow spores shed water and have been used as baby powder.

DESCRIPTION: Low, evergreen, perennial herbs with erect, leafy, bottlebrush-like stems 2–12" (5–30 cm) tall. Leaves firm, shiny, needle-like, $1/8$–$3/8$" (3–10 mm) long, in 8 dense, vertical rows. Reproducing by spores.

WARNING:
Running clubmoss and mountain fir clubmoss contain toxic alkaloids. Clubmoss spores can irritate mucous membranes and sensitive or damaged skin. People exposed to large amounts of these spores (e.g., in factories where spores are used in dusting powder) have developed asthma.

Stiff clubmoss (*L. annotinum*) has pointed leaves on unbranched to twice-forked shoots from creeping, irregularly rooted stems up to 3¹/₄' (1 m) long. Its spores are borne in single, stalkless, ¹/₂–1³/₈" (12–35 mm) long cones (strobili) at the stem tips. It grows in moist foothill, montane and subalpine sites from Alaska to Colorado.

Running clubmoss (*L. clavatum*) is very similar to stiff clubmoss, but its young leaves are tipped with a slender, hair-like bristle, and its fertile stems have 2 or more strobili on a long stalk. It grows in similar habitats from Alaska to Montana.

Mountain fir clubmoss (*L. selago* or *Huperzia haleakalae*) and **northern fir clubmoss** (*H. selago*) are tufted species about 2–4" (5–10 cm) tall, with spore clusters in the axils of upper leaves (rather than in cones). Both grow on moist to dry slopes in alpine and subalpine zones, mountain fir clubmoss from Alaska to Montana and northern fir clubmoss from the Yukon to BC.

Stiff clubmoss

Running clubmoss

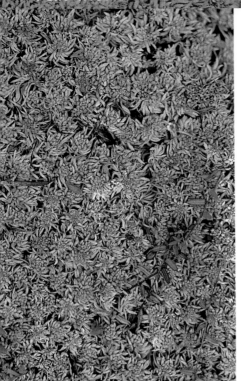
Peat moss

Peat mosses
Sphagnum spp.

FOOD: Peat mosses have rarely been used for food, though they were included in a list of famine foods from China and as an ingredient for breads in Lapland. Generally, peat moss was considered a 'wretched food of barbarous countries' (Theiret 1956).

MEDICINE: These extremely absorbent plants are permeated with minute tubes and air spaces that act as a fine sponge. Because peat mosses grow in acidic environments, they also have astringent and antiseptic properties. Even the air around peat bogs was believed to be healthful, owing to the absorption of hydrogen and exhalation of oxygen by the mosses. Peat moss was used in surgical dressings for many years and was very important in this capacity during World War I, freeing cotton for the production of gunpowder. It absorbs liquids 3 times as quickly as cotton and holds 2–4 times the volume. Compared to cotton, peat-moss dressings required less frequent changing, distributed liquids more uniformly throughout the mass and were cheaper, cooler, softer and less irritating. Swellings, broken bones, dislocations and serious cuts and infections were wrapped in heated peat moss or held over steam and then covered with peat moss and a blanket. Sore eyes were treated with hot moss compresses, and sore ears were steamed over heated peat moss. Red peat moss, soaked in cold water, was placed on the head to cure headaches and on the chest to treat lung problems. Peat moss was boiled in water to make medicinal teas for treating bleeding and eye diseases. It was also mixed with grease to make salves for cuts. Peat-moss dressings and sphagnol (a distillate of peat tar) have been used to treat many skin problems, including eczema, psoriasis, hemorrhoids, scabies, acne and insect bites.

OTHER USES: Peat moss provided soft mattresses, blankets and disposable diapers that kept infants remarkably clean, dry and warm. It was also used as toilet paper and menstrual padding. Mattresses and quilts were stuffed with peat moss, and drafty walls and leaky boats were caulked with it. Peat-moss smudges repelled flies and mosquitoes in summer and protected against frost in winter. The compacted lower layers in bogs (peat) have been used as fuel for centuries. Other uses of peat include production of ethyl alcohol, acetic acid, carbonic acid, ammonium compounds, nitrates, paraffin, naphtha, pitch, charcoal, brown dye, tanning materials, lignins for making plastics, paper, shoe insoles, woven fabrics, artificial wood, clay for porous bricks and stuffing

WARNING:
Spores from fungi in peat moss have caused lung infections (sporotrichosis) in nursery workers and people harvesting peat moss.

for life preservers. Peat moss has also been used as an absorbent, deodorizing animal bedding that controls insect pests. It is used extensively as a potting medium and as packing material for perishable items—helping to retain essential moisture for long periods and to insulate against heat and cold. Peat moss has even been mixed with molasses and used for stock feed and mixed with lard as dog food in times of food shortage.

DESCRIPTION: Large mosses, often forming tussocks, mats and lawns. Stems erect, with distinct heads of crowded branches (capitula) at their tips and clusters of branches (fascicles) along their length. Leaves concave, with a net-like pattern of slender, green, living cells and large, clear, dead cells. Branch leaves crowded and overlapping. Stem leaves relatively large and scattered. Reproducing by spores in

Warnstorf's peat moss

round, black capsules on short, clear stalks. Peat mosses are common in bogs throughout the Rocky Mountains, but a few species (e.g., Warnstorf's peat moss) grow in fens. The 4 species below are common in the northern and Canadian Rockies.

Warnstorf's peat moss (*S. warnstorfii*) has relatively slender, purple-tinged plants with flat heads and with leaves in 5 vertical rows.

Midway peat moss (*S. magellanicum*) has fat, swollen, red plants, **yellow-green peat moss** (*S. angustifolium*) has yellow-green, bushy plants, and **rusty peat moss** (*S. fuscum*) is rusty brown.

Iceland-lichen

Iceland-lichen

Cetraria islandica & C. ericetorum

ALSO CALLED: Iceland-moss.

FOOD: Lichens have very little protein and fat and only trace amounts of vitamins and minerals, and their carbohydrates (they are about 80 percent starch) are only partially digestible by humans. Consequently, their main food value appears to be in providing bulk. Iceland-lichen is among the most highly rated of the edible lichens. Bitterness and astringency were reduced by adding a spoonful of baking soda to the cooking water. Fresh, or dried and powdered, lichens were boiled in milk or water (1 part lichen to 3–4 parts water) to make gruel (hot), jelly (cooled and allowed to set) or thickener for soups and stews. Lichen gruel was flavored with salt, raisins or cinnamon, and the jelly was often mixed with lemon juice, sugar, chocolate or almonds. In bread-making, lichen was boiled with lye, rinsed with cold water, dried and ground into flour, which was mixed 3:1 with grain meal and baked. This mixture produced a strong, fair-tasting bread that stored well and was less likely to be attacked by insects. Equal parts of lichen and wheat flour were used to make muffins. Lichen jelly could also be sliced, dried and ground into flour for bread-making.

MEDICINE: Iceland-lichen is rich in mucilage (40–50 percent), and it has long been used as a remedy for coughs and colds. It is still sold in Scandinavia for this purpose. An ounce of lichen boiled in 2 cups of water was said to ease coughing and inflammation of the mucous membranes in colds, bronchitis and tuberculosis. Some herbalists considered Iceland-lichen an important blood purifier and bitter spring tonic for restoring, invigorating, refreshing and stimulating the system. Plants were soaked overnight in cool water to make a bitter medicinal tea for controlling diarrhea (though some sources classify this lichen as a laxative), aiding digestion and improving appetite. Iceland-lichen was also used as a dietary supplement for convalescents and people suffering from general weakness. Lichen substances, such as usnic, physodic, evernic and vulpinic acids, are relatively non-toxic and have proved effective against a wide range of bacteria (including a penicillin-resistant strain of *Staphylococcus aurea*), protozoa (including *Trichomonas*) and fungi (including athlete's foot and ringworm). They have been used in the treatment of ulcers, burns, plastic surgery and various skin problems.

WARNING:
Edible lichens are often difficult to distinguish from inedible ones. Iceland-lichen can cause severe digestive upset. Soaking in water for several hours, draining and adding baking soda or ashes to the cooking water help to remove irritating compounds. Lichens grow very slowly (only a few millimeters per year) and can take many decades to regrow, so they are very sensitive to disturbance.

About 88 lb (40 kg) of Iceland-lichen was required to make 22.2 lb (1 kg) of antibiotic. Powdered lichen was used to dust mouth sores. Compounds isolated from Iceland-lichen have been found to both stimulate the immune system and inhibit one of the enzymes essential to replication of HIV. These compounds are being studied as a potential alternative to toxic drugs such as AZT.

OTHER USES: Iceland-lichen was widely used as a brown dye in the days before synthetic dyes. It was also fermented by distillers to produce alcohol. These lichens provide forage for reindeer and were sometimes fed to cattle and horses. Water in which Iceland-lichen has been soaked contains the enzyme ribonuclease, which destroys the nucleic acids of tobacco mosaic virus. Even when diluted 1:500, this enzyme inhibits the development of the virus by about 80 percent.

C. islandica (both photos)

DESCRIPTION: Medium-sized, upright leaf lichen, ³/₄–2" (2–5 cm) tall, with strongly incurved, often minutely spiny edges. Brownish on both surfaces, with white breathing pores (pseudocyphellae) on the lower side, forming a white line near the lobe edges (*C. ericetorum*) or scattered over the surface in white patches (*C. islandica*). Occasionally reproducing by spores in brown, disc-like structures (apothecia) at the lobe tips. Iceland-lichen grows on moist, open, well-drained sites in montane, subalpine and alpine zones from the Yukon to New Mexico.

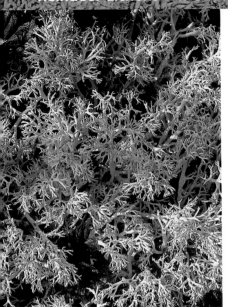

Green reindeer-lichen

Reindeer-lichens

Cladina spp.

ALSO CALLED: reindeer-moss.

FOOD: Reindeer-lichens have been dried, ground into flour and used to thicken soups, sauces and desserts, or mixed with other flour to make breads and muffins. Bitterness can be reduced by boiling with a spoonful of baking soda in the cooking water. Because lichen carbohydrates cannot be digested completely by humans, these plants are most nutritious after they have been partially digested by a ruminant. Northern native peoples often ate the half-digested, lichen-rich contents of caribou stomachs. The lichen was separated from the grass and then eaten with oil, or it was mixed with blood and fat, meat scraps or liver and cooked to make a savory pudding. Dried, pulverized lichens were boiled or soaked in hot water until soft and eaten plain or, preferably, mixed with berries, fish eggs or grease. Reindeer-lichens were also boiled with caribou blood for food. The word *teniyash*, which means 'increase,' was sung while stirring the lichen mixture, so that it would rise and become light. In the 1800s, a Swedish distillery used gray reindeer-lichens to make a popular brandy, but this practice ended after large areas had been stripped of lichens and supplies ran out.

MEDICINE: Gray reindeer-lichen was boiled to make medicinal teas for treating fevers, diarrhea, jaundice, tuberculosis, convulsions and other diseases. In Finland, it was boiled in water and taken as a laxative, or it was boiled in milk and taken for coughs. Usnic acid, extracted from these lichens, has been used in healing salves.

OTHER USES: In the subarctic, reindeer-lichens provide critical winter food for caribou, but in the Rockies, heavy snow buries these plants too deeply and caribou are forced to rely on tree-dwelling species, such as hair lichens (pp. 232–33). Gray reindeer-lichen is one of the most common species grazed by northern caribou, and for 6 months of the year, it is an important food for domestic reindeer herds in northern Europe. A reindeer needs the equivalent of 4 1/2 lb (2 kg) of dry lichen daily, which requires grazing about 14 sq yd (12 sq m) to a depth of 1 1/4" (3 cm). Grazed lichens re-grow at a rate of about 3/16" (about 4 mm) a year. Therefore, they cannot be cropped for 2–5 years if grazing is modest, or for 10–15 years if it is intense. These lichens were sometimes collected as fodder for cattle and other domestic animals, including dogs. It was said that the milk of cattle fed with reindeer-lichen was creamier and the meat was fatter and sweeter.

WARNING:
Acids in reindeer-lichens may cause stomach upset, especially when the lichens are not cooked well. A few people are sensitive to lichen substances (eg., usnic acid) and develop lichen dermatitis, which is characterized by itchy, reddened skin, sometimes accompanied by pustular rashes or scaliness.

DESCRIPTION: Upright shrub lichens, about 2–3" (5–8 cm) tall, with hollow, intricately branched stems covered in dull, cottony hairs (use a hand lens), often forming extensive carpets. Occasionally reproducing by spores borne in round, brown structures (apothecia) at the branch tips.

Green reindeer-lichen (*C. mitis* and *C. arbuscula*) has pale yellowish-green branches from a distinct main stem. These 2 species are distinguished mainly by their different chemistry (revealed by chemical tests).

Gray reindeer-lichen (*C. rangiferina*), as the common name suggests, is pale gray.

Reindeer-lichens grow on open or wooded sites in foothills, montane, subalpine and alpine zones from Alaska to Colorado.

Gray reindeer-lichen

Green reindeer-lichen (left) and gray reindeer-lichen (right)

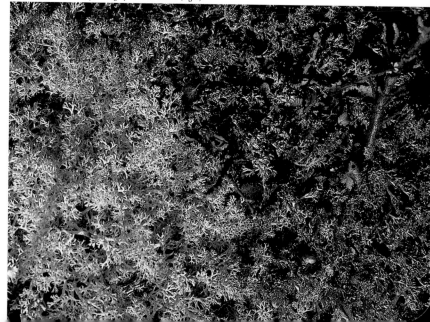

Hair lichens

Alectoria spp.,
Bryoria spp. & *Usnea* spp.

Speckled horsehair-lichen

FOOD: Edibility and palatability of hair lichens vary greatly with the species (see Warning) and even with the type of tree on which they are growing. Large volumes of lichen were gathered either by using long poles or by climbing or felling trees. Five or six lichen-covered trees could supply a family for a year. Raw lichens could be eaten, but they were often quite bitter. Some bitterness was removed by lengthy soaking in fresh water, by thorough cooking, and by adding baking soda or ashes to the cooking water. Hair lichens were usually pit-steamed for 12–48 hours, with wild onions, roots and bulbs added for flavoring. An 8" (20 cm) layer of lichen produced a black, gelatinous mass 1¹/₂–2" (4–5 cm) thick. More recently, lichens have been boiled in heavy saucepans for several hours (or in pressure cookers for shorter periods), and sugar or apples have been used for sweetener, but these present-day methods are said to produce an inferior product. Cooked lichen was cooled, sliced and eaten with fish, berries or grease, or it was dried for later use. Dried cakes were boiled with berries, roots or meat, and thin, dried slices were dipped in soup, like crackers. Fresh or cooked lichen was sometimes dried and ground into flour for thickening soups, stews and cereals and for adding to breads. Occasionally, edible horse-hair lichen was roasted until crumbly and then boiled until it was like molasses. Old-man's beard was sometimes boiled and eaten with fish, berries or grease. Because these lichens grow on trees, they can provide a useful emergency food year-round.

MEDICINE: In the Middle Ages, old-man's beard was used to treat lung diseases and (based on the Doctrine of Signatures) scalp disorders. In Scandinavia, it was boiled to make a wash for bathing chapped skin on babies or the feet of adults. Old-man's beard contains the antibiotic substance, usnic acid, which has been found to be effective against the bacteria that causes tuberculosis (*Mycobacterium tuberculosis*). It is still used in China to treat this disease. It has also been used to treat inflamed mucous membranes (catarrh), water retention (edema) and whooping cough. Extracts from these lichens are effective against a wide spectrum of bacteria and fungi. They may be even more

WARNING:
Lichen acids, such as usnic acid and vulpinic acid, make many lichens bitter and even poisonous. Experienced collectors tasted the lichens in an area before collecting quantities, to ensure that the right species was gathered.

effective than penicillin for inhibiting the growth of gram-positive bacteria (e.g., *Streptococcus* spp., *Pneumococcus* spp.). Old-man's beard also stimulates the immune system. In Europe, usnic acid from lichens is used in commercial ointments for treating skin problems ranging from fungal infections to burns. Old-man's beard can provide an emergency dressing to stop bleeding and prevent infection.

OTHER USES: Witch's-hair produces a yellow dye, and it has also been fermented by distillers to produce alcohol. Occasionally, it was used to make capes, blankets and shoes, mainly for poor people who had no animal skins for clothing. Hair lichens are an important source of winter food for ungulates, and consequently they were very important, both directly and indirectly, to the winter food supply of many tribes. Lichens are very sensitive to sulphur dioxide and other pollutants, so in many areas they are used to monitor air quality. Most lichens quickly disappear from polluted areas.

DESCRIPTION: Tufted, hair-like lichens, usually hanging on trees. Reproducing mainly by fragmentation and by small, powdery propagules from white dots (soredia).

One of the most common, **speckled horsehair-lichen** (*B. fuscescens*) has shiny, brown to blackish, pliant branches, with scattered, white, dot-like soredia.

Edible horsehair-lichen (*B. fremontii*) has shiny, grooved, variably thickened, mostly yellowish-brown branches about 12–18" (30–45 cm) long.

Old-man's beard (*Usnea* spp.) includes many species of pale yellowish-green, highly branched hair lichens, with branches that are reinforced by a strong, rather elastic central cord.

Witch's-hair (*Alectoria* spp.) resembles old-man's beard, but its branches lack a strong central cord.

Hair lichens grow in the foothills, montane and subalpine zones from Alaska to New Mexico.

Bryoria spp. (top); *Usnea* (above)

233

Poisonous Plants

Baneberry, with white berries

Thiis section describes 46 of the more common toxic plants in the Rocky Mountains, but it is by no means a complete guide to poisonous plants of the region. Many potentially toxic species in this book are not found in this section. Some have been widely used by humans and are discussed in the main body of the book, with notes about toxicity under the heading 'Warning.' For example, many of the drugs produced by plants can be deadly poisons when improperly prepared or administered.

Plants have developed a wide array of protective strategies to compete against other plants and to avoid infection and predation (being eaten or otherwise used) by animals. Many of these protective mechanisms are dangerous to humans, but as with most natural systems, the effects can vary greatly from plant to plant and from person to person. There are very few deadly poisonous plants in the Rockies. In most cases, you would have to eat large quantities to be fatally poisoned, but even though these plants may not kill you, you may wish you were dead when you experience their effects.

Toxins are rarely distributed evenly throughout all parts of a plant. For example, you may enjoy eating cherries, but did you know that all parts of these trees and shrubs, with the exception of the flesh of the cherries, contain hydrocyanic acid, which causes cyanide poisoning? Many of us enjoy rhubarb, but only the leaf stalks are edible. The broad, green leaf blades contain toxic concentrations of oxalates. Toxicity often changes as plants grow. Some species are most poisonous when they are young, whereas others grow increasingly toxic with age, often concentrating poisons in their seeds.

The toxicity of most wild plants has yet to be studied, but many are known to be poisonous. People can be harmed by plants in many ways.

1. Ingestion of plant toxins

Plant toxins are generally most effective when they are taken into the body in plants or plant extracts (e.g., as teas or tinctures) and absorbed into the system through the digestive tract. Types of toxins in this group include

(a) alkaloids: bitter-tasting chemicals that tend to be mildly alkaline. Many alkaloids affect the nervous system. This group includes some of our most powerful drugs (e.g., morphine, mescaline, caffeine, nicotine and strychnine).

(b) glycosides: two-parted molecules composed of a sugar such as glucose and a non-sugar or aglycone. Usually, glycosides become poisonous when they are digested and the sugar is separated from its poisonous

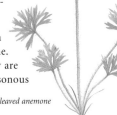

Cut-leaved anemone

aglycone. Some of the more common aglycones include cyanide, cardioactive steroids (which affect the heart), saponin and mustard oil (which irritate the digestive tract) and coumarin (a powerful anti-coagulant).

(c) organic acids: carbon-based acids that lack nitrogen. Our only common organic acid, oxalic acid, causes ionic imbalances, internal bleeding and kidney damage. It is present in many plants, but seldom in toxic concentrations.

(d) alcohols: all alcohols are somewhat toxic, but some are more poisonous than others. The deadly poison of water-hemlock (p. 241), cicutoxin, is an unsaturated alcohol that causes convulsions and death in a few minutes to a few hours.

(e) resins and resinoids: a varied group of chemicals, many of which are phenolic compounds. Some phenolic resins (e.g., tetrahydrocannabinol of marijuana [*Cannabis sativa*]) affect the nervous system. Others cause skin reactions. Urushiol in poison-ivy (p. 237) can cause rashes and blistering, and hypericin in St. John's-wort (p. 141) can cause reactions when skin is exposed to sunlight.

Baneberry, with red berries

2. Ingestion of bacteria and other parasites in or on plants

Not all poisonings result from substances produced by the plants. Bacteria, protozoa, nematodes and other animals growing in soil or water can adhere to plants, and they will cause parasitic infections when plants are eaten raw without proper cleaning. This situation is most common with aquatic plants growing in polluted water. Some fungi that infect plants can also be toxic to humans. One of the best-known examples of toxic fungi is ergot (*Claviceps*), a group of parasitic fungi that infects the grains of grasses (pp. 216–18) and causes ergotism when eaten by humans.

3. Ingestion of chemical pollutants in or on plants

Soil, water and air can become polluted with chemicals from factories (often widespread) or vehicle exhaust (along roadsides) or with herbicides and pesticides (in parks and school yards and along roads). These compounds may be absorbed by plants or may simply coat the plant stems and leaves. Many pollutants are not removed by washing and are ingested when plants from polluted environments are eaten. Not all harmful chemicals are synthetic. For example, when selenium is present in the soil, some plants (e.g., paintbrushes [*Castilleja* spp.]) take it up and concentrate it to toxic levels in their tissues, making their plants poisonous. Similarly, many plants can accumulate toxic amounts of nitrates or molybdenum.

WHAT TO DO

Proper treatment for cases of poisoning requires more information than this guide provides. Usually, it is best to induce vomiting, but in some cases vomiting can do more harm than good. If a person is believed to have been poisoned, consult a doctor. If you suspect that a certain plant (or plants) is responsible for the problem but do not know the plant's identity, take samples (preferably whole plants, with roots and flowers or fruits) that can be examined later for positive identification.

Yellow columbine

235

4. Contact with irritating plants

Human skin is sensitive to contact with many plants that do not affect other animals. Sensitivity varies greatly from person to person. For example, most people develop rashes from contact with poison-ivy, but a few can handle these plants with impunity. Skin reactions may be caused by plant toxins or may simply result from irritation from protective hairs and spines.

Many plants (e.g., lady's-slipper orchids [*Cypripedium* spp.] and scorpionweeds [*Phacelia* spp.]) are covered with stiff, protective hairs that irritate the skin when the plant is touched or eaten. These hairs cause allergic reactions (dermatitis) in some people, but they have no effect on others. A few species have developed specialized hairs to irritate potential grazers. The hairs of stinging nettles (pp. 200–01) act like miniature hypodermic syringes, injecting droplets of acid under the skin and causing a burning sensation in everyone.

False azalea

Stiff bristles, spines and thorns are obvious protective mechanisms that can damage skin. Some have barbs that help them to catch onto passers-by and to work their way into tender tissues around the mouth and eyes of grazing animals. These structures can also cause serious wounds and obstructions in the digestive tract.

A few plants produce irritating resins that stick to hair and clothing and cause rashes and blisters when they come into contact with skin. Poison-ivy (p. 237) is one of the best-known plants in this group. Another group of compounds (e.g., furanocoumarins in cow-parsnip [p. 191]) causes dark blotches, rashes and even blisters when plants are eaten and the skin is then exposed to light.

5. Contact with plant allergens

Allergy-producing materials are life-threatening to relatively few people, but in these cases their effects can be as dangerous as those of any toxin. Allergic reactions vary greatly from person to person and plant to plant. Air-borne spores and pollen go unnoticed by most, but in people with allergies, specific types of pollen and spores can cause anything from a runny nose to a severe asthma attack. Similarly, foods that most of us enjoy everyday can cause fatal reactions in people who have developed allergies to a specific ingredient (e.g., peanuts). Most skin reactions (including poison-ivy rashes) are allergic reactions.

Common snowberry

Poison-ivy

Toxicodendron radicans

Poison-ivy plants contain an oily resin (urushiol) that causes a nasty skin reaction in most people, especially on sensitive skin and mucous membranes. Urushiol is not volatile and therefore it is not transmitted through the air, but it can be carried to unsuspecting victims on pets, clothing and tools and even on smoke particles from burning poison-ivy plants. The resin can be removed by washing with a strong soap. Washing can prevent a reaction if it is done shortly after contact, and it also prevents transfer of the resin to other parts of the body or to other people. The liquid that oozes from poison-ivy blisters on skin does not contain the allergen. The Navajo rubbed sheep blood over affected skin. Ointments and even household ammonia can be used to relieve the itching of mild cases, but people with severe reactions might need to consult a doctor.

DESCRIPTION: Trailing to erect deciduous shrub, about 4–10" (10–25 cm) tall, forming colonies. Leaves bright glossy green, divided into 3 pointed, oval leaflets, scarlet in autumn. Flowers cream-colored, 5-petaled, $^1/_8$" (4–5 mm) across, forming crowded clusters, in May to July. Fruits whitish, berry-like drupes $^1/_4$" (5–7 mm) wide. Poison-ivy grows on well-drained plains and foothill sites from BC and Alberta to New Mexico.

Small bog-laurel

Kalmia microphylla

ALSO CALLED: swamp-laurel.

These plants contain alkaloids called 'andromedotoxins,' which can cause watering of the mouth, nose and eyes, headaches, vomiting, weakness and eventually paralysis, when consumed in sufficient quantities. Symptoms may appear 3–14 hours or more after the plant has been eaten. Poisoning and death of cattle, sheep, goats and horses have been reported. Even the honey made from bog-laurel nectar is said to be poisonous.

DESCRIPTION: Slender evergreen shrub, 2–8" (5–20 cm) tall. Leaves leathery, glossy green, opposite, oval to elliptic-oblong, with down-rolled edges and a whitish lower surface. Flowers rose-pink, saucer-shaped, about $^1/_4$–$^3/_8$" (5–10 mm) across, forming loose clusters at the branch tips, in June to September. Fruits round capsules. Small bog-laurel grows in moist to wet, open sites in subalpine and alpine zones from the Yukon and NWT to Colorado.

False azalea

Menziesia ferruginea

Although these plants are less toxic than many other members of the Heath family, they do contain andromedotoxins. These alkaloids can cause watering of the mouth, nose and eyes, headaches, vomiting, weakness and eventually paralysis, when consumed in sufficient quantities. Poisonings and deaths of sheep, as a result of eating false azalea, have been reported. Symptoms include weakness, salivation, vomiting, difficulty breathing and paralysis.

DESCRIPTION: Erect, sometimes skunky-smelling deciduous shrub with fine, rust-colored, glandular-sticky twigs. Leaves dull blue-green, elliptic, with the midvein protruding at the tip, broadest above the middle, $1^1/4$–$2^3/8$" (3–6 cm) long, crimson-orange in autumn. Flowers peach-colored, urn-shaped, $1/4$" (6–8 mm) long, nodding in small, loose clusters, in May to July. Fruits oval capsules $1/4$" (5–7 mm) long. False azalea grows on moist, wooded slopes in foothills and montane zones from BC and Alberta to Wyoming.

White rhododendron

Rhododendron albiflorum

Rhododendrons contain alkaloids called andromedotoxins, which can cause watering of the mouth, nose and eyes, headaches, vomiting, weakness and eventually paralysis and death, when consumed in sufficient quantities. Children have been poisoned by eating the leaves and flowers of white rhododendron, and sheep and occasionally cattle have died from eating these bitter shrubs.

DESCRIPTION: Deciduous shrub, $1^1/2$–$6^1/2$' (50–200 cm) tall, with slender, reddish-hairy branches. Leaves alternate, crowded near the branch tips, glossy green above but pale beneath, narrowly elliptic to lance-shaped, $3/4$–$3^1/2$" (2–9 cm) long, with the midvein not protruding from the tip. Flowers white, cup-shaped, 5-petaled, $5/8$–$3/4$" (1.5–2 cm) across, nodding in clusters of 1–4, in June to August. Fruits woody, oval, glandular-hairy capsules $1/4$" (6–8 mm) long. White rhododendron grows in moist forests in montane and subalpine zones in BC, Alberta and Montana.

Common snowberry

Symphoricarpos albus

Some sources report that these berries are edible, though not very good. However, snowberries are mildly poisonous and large quantities can be toxic. The branches, leaves and roots are also poisonous. These plants contain the alkaloid chelidonine, which can cause vomiting, diarrhea, depression and sedation. One child who ate the berries reportedly became nauseous, delirious and fell into a semi-comatose state. Most tribes considered snowberries poisonous. Some believed they were the ghosts of saskatoons, part of the spirit world and not to be eaten by the living.

DESCRIPTION: Erect deciduous shrub, usually 20–30" (50–75 cm) tall, with pale green, opposite, elliptic to oval leaves ³/₄–1¹/₂" (2–4 cm) long. Flowers pink to white, broadly funnel-shaped, ¹/₈–¹/₄" (4–7 mm) long, borne in small clusters at the stem tips, in June to August. Fruits white, waxy, berry-like drupes ¹/₄–³/₈" (6–10 mm) long, persisting through winter. Common snowberry grows on well-drained sites in plains to lower subalpine zones from BC and Alberta to Colorado.

Seaside arrow-grass

Triglochin maritima

Arrow-grasses contain 2 glycosides (triglochinin and taxiphillin) that release hydrocyanic acid when the plants are chewed. These compounds are most concentrated in young flowering stalks. Also, water-stressed plants can be 5–10 times more poisonous than plants growing in water. Dried plants remain poisonous. Doses of only 0.5 percent of body weight can be lethal. Ruminants are most commonly poisoned. Many cattle and sheep have died from eating seaside arrow-grass. Symptoms include nervousness, trembling, salivation, vomiting, convulsions and death from respiratory failure. Some tribes gathered the sweet, white stem bases of these plants (the green upper parts were discarded) and ate them raw, cooked or canned. The seeds were also parched and eaten or used as a coffee substitute. The seeds and stem bases have relatively low glycoside concentrations, but arrow-grasses should be used with caution, if at all.

DESCRIPTION: Clumped, perennial herb with fleshy, basal, grass-like leaves about ¹/₁₆" (2 mm) wide. Flowers inconspicuous, ¹/₈" (3–4 mm) long, 6-petaled, greenish, with feathery, red stigmas, forming 4–16" (10–40 cm) long spikes, in May to August. Fruits oval capsules, about ¹/₄" (5 mm) long. Seaside arrow-grass grows on wet, open sites in plains, foothills and montane zones from Alaska to New Mexico.

Mountain death-camas

Death-camases
Zigadenus spp.

All parts of these plants contain poisonous steroidal alkaloids (e.g., zygacine, zygadenine) that are said to be more potent than strychnine. Two bulbs, raw or cooked, can be fatal. Severe poisoning causes digestive upset and vomiting followed by loss of muscle control, lowered blood pressure and temperature, difficulty breathing and eventually coma and death. If someone has eaten this plant, induce vomiting and get medical help. Sheep (herds of as many as 500), cattle and horses have died from eating these plants, usually in spring when little other forage is available. Meadow death-camas is about 7 times more poisonous than mountain death-camas, but both are dangerous.

DESCRIPTION: Gray-green, perennial herbs, 6–28" (15–70 cm) tall, with grass-like leaves from oval, blackish-scaly bulbs. Flowers whitish, 6-petaled, with 6 glands near the center, forming elongated clusters, in June to August. Fruits erect, 3-lobed capsules.

Mountain death-camas (*Z. elegans*) has ³/₄" (2 cm) wide flowers, with green, heart-shaped glands. It grows on moist slopes in foothills to alpine zones from Alaska to New Mexico.

Meadow death-camas (*Z. venenosus*) has ¹/₂–⁵/₈" (1–1.5 cm) wide flowers, with round to oblong glands. It grows at lower elevations from BC and Alberta to Colorado.

Green false-hellebore
Veratrum viride

False-hellebores are violently poisonous and can cause birth defects in pregnant animals. The rootstocks, roots and young shoots are the most toxic. They contain steroidal alkaloids that slow heartbeat and breathing and lower blood pressure. Cardiac and respiratory stimulants (e.g., atropine) are used in treatment. Symptoms include frothing at the mouth, nausea, blurred vision, lockjaw, vomiting and diarrhea. People have reported stomach cramps after drinking water in which this plant was growing. Cattle, sheep, poultry and mice have been poisoned by false-hellebores. Powdered false-hellebore has been used in the garden insecticide 'hellebore' and in salves and shampoos for treating scabies, herpes, ringworm, lice and mites. Some native peoples used these plants to poison arrows and to commit suicide.

DESCRIPTION: Robust, perennial herb with leafy, unbranched stems 2¹/₄–6¹/₂" (70–200 cm) tall. Leaves alternate, clasping, elliptic, 4–10" (10–25 cm) long, prominently parallel-veined or accordion-pleated. Flowers green, 6-petaled, about ³/₄" (2 cm) across, musky-smelling, forming open clusters 12–28" (30–70 cm) long, with tassel-like branches, in June to September. Fruits straw-colored to brown, ovoid capsules ³/₄–1¹/₄" (2–3 cm) long.

Western blue flag

Iris missouriensis

Blue flag rootstocks, roots and young shoots are toxic and should never be taken internally, but their sharp, bitter taste usually prevents the consumption of sufficient quantities to cause poisoning. Irises contain the acrid, resinous substance irisin, which irritates the digestive tract, liver and pancreas. Symptoms include stomach upset with a burning sensation, difficulty breathing, vomiting and diarrhea. Some people develop severe allergic skin reactions from handling iris plants and iris rootstocks in particular. Some native peoples placed a mashed piece of western blue flag rootstock on tooth cavities or gums to kill the pain.

DESCRIPTION: Clumped, perennial herb, about 8–20" (20–50 cm) tall, with sword-shaped leaves $^3/_{16}$–$^3/_8$" (5–10 mm) wide. Flowers pale blue to deep blue, about 2$^1/_2$" (6–7 cm) across, with 3 backward-curved, purple-lined sepals, 3 erect, narrower petals and 3 flattened, petal-like style branches, borne in groups of 2–4, in May to July. Fruits 1$^1/_8$–2" (3–5 cm) long capsules. Western blue flag grows on wet (at least in spring) sites in plains, foothills and montane zones from BC and Alberta to New Mexico.

Water-hemlocks

Cicuta spp.

All parts of these plants contain the resin-like toxin, cicutoxin, but the rootstocks are the most poisonous. Small amounts can be deadly: 2–3 bites for humans; 3$^1/_2$ oz (100 g) for sheep; 11 oz (300 g) for horses; and 14 oz (400 g) for cattle. Even children making pea-shooters and whistles from the hollow stems have been poisoned. The poison acts on the central nervous system, and it can take effect within 15 minutes of ingestion. Symptoms include stomachache, salivation, nausea, vomiting, diarrhea, difficulty breathing, tremors and violent convulsions. If a person has eaten water-hemlock, induce vomiting, administer a strong laxative and consult a doctor.

Douglas' water-hemlock

DESCRIPTION: Perennial herbs with foul-smelling, oily, yellow sap and thickened, chambered stem bases and/or roots. Leaves alternate, 2–3 times divided into leaflets. Flowers white to greenish, tiny, forming twice-branched, flat-topped clusters (with few or no bracts), in June to August. Fruits flattened, round to oval seeds, with thick, corky ribs.

Bulbous water-hemlock (*C. bulbifera*) is 1–2$^1/_2$" (30–80 cm) tall with linear, $^1/_{32}$–$^3/_{16}$" (0.5–4 mm) wide leaflets and with small bulblets in its leaf axils. It grows in wet sites from the Yukon and NWT to Montana.

Douglas' water-hemlock (*C. douglasii*) is 2–6' (60–180 cm) tall and has lance-shaped to oblong leaflets whose veins end at the tooth bases (remember, vein to the cut, pain in the gut). It grows on wet sites in plains to subalpine zones from the Yukon and NWT to New Mexico.

Poison-hemlock
Conium maculatum

All parts of these plants are extremely poisonous, but the toxic alkaloids are most abundant in the young plants, leaves and flowers. These toxins affect the nervous system, causing numbness and paralysis of the lower limbs, followed by paralysis of the arms and chest. Symptoms include nervousness and confusion, weakness, vomiting and diarrhea, weak pulse, difficulty breathing and finally death from suffocation. If a person has eaten this plant, induce vomiting, administer a strong laxative, and consult a physician. Fatal doses vary with the specific animal or person. Lower levels can produce birth defects. Some people develop a rash from contact with poison-hemlock, so wash your hands carefully if you must handle these plants. These plants have been mistaken for wild carrot, anise or parsley, with fatal results.

DESCRIPTION: Musty-smelling, hairless, biennial herb with branched, leafy, purple-blotched stems 1¹/₂–10' (50–300 cm) tall. Leaves lacy and fern-like, 6–12" (15–30 cm) long. Flowers white, tiny in twice-branched, flat-topped clusters above a whorl of small, lance-shaped bracts. Fruits oval, slightly flattened, about ¹/₁₆" (2 mm) long, with raised, often wavy ribs. This European species has spread to disturbed sites across North America.

Anemones
Anemone spp.

All anemones are somewhat poisonous, but toxicity varies greatly with habitat and between species. The sap contains the glycoside ranunculin (see buttercups, p. 243). Small amounts of mashed anemone leaves were sometimes used as counter-irritants for treating bruises and sore muscles. Protoanemonin is volatile, so dried or thoroughly cooked plants are said to be harmless.

DESCRIPTION: Perennial herbs with alternate, divided leaves. Flowers usually white, with 5 petal-like sepals and no petals. Fruits tiny, beaked seed-like achenes in dense heads.

Cut-leaved anemone (*A. multifida*) is 8–20" (20–50 cm) tall, with finely divided, palmately lobed leaves, wooly achenes and 1 to several, white to reddish flowers above a whorl of stem leaves. It grows on dry slopes in foothills, montane and subalpine zones from Alaska to New Mexico.

Long-headed anemone (*A. cylindrica*) has long-stalked leaves and cylindrical seed heads. It grows in open, low-elevation sites from BC and Alberta to New Mexico.

Cut-leaved anemone

Buttercups

Ranunculus spp.

All buttercups are somewhat poisonous, but toxicity varies greatly with habitat and species. Meadow buttercup and cursed buttercup are among the most toxic. Buttercup sap contains the glycoside ranunculin, which is converted into an irritating yellow oil, protoanemonin, when the plant is damaged. Protoanemonin can cause intense pain and burning of mucous membranes (e.g., in the digestive tract) and may raise blisters on sensitive skin. Glycoside concentrations are highest during flowering. Fresh buttercups blister the mouths of grazing animals and can cause salivation, abdominal pains, diarrhea, slow heartbeat, muscle spasms, blindness and, rarely, death. Buttercups also have a narcotic effect on cattle and give a bitter taste to their milk. Human poisonings are rare (probably because the most toxic plants are so acrid), and they usually result only in digestive upset. In the past, meadow buttercup juice was used as a counter-irritant for treating rheumatism, arthritis and neuralgia (severe pain along a nerve), and it was applied to blister or eat away warts, pimples and plague sores. Beggars sometimes used it to raise blisters and gain sympathy from passers-by. The English believed that the smell of buttercup flowers could drive a person mad and induce epileptic seizures. In first century A.D., the Roman scholar Pliny warned that if people ate *R. illyricus* they would burst into gales of laughter, ending in death. The only antidote was a mixture of pineapple kernels and pepper dissolved in wine. Protoanemonin is volatile, so dried or thoroughly cooked plants are said to be harmless. The dried, ground fruits and cooked roots and leaves of some buttercups have been used for food, but this practice is not recommended.

Cursed buttercup (top); Meadow buttercup (above)

DESCRIPTION: Low, perennial herbs with alternate, variously divided leaves. Flowers usually yellow, with 5 sepals and 5 petals; each petal with a nectar-bearing spot at the base. Fruits tiny seed-like achenes in dense, round to cylindrical heads.

Meadow or **tall buttercup** (*R. acris*) is up to 12–40" (30–100 cm) tall and has palmately 3–5-lobed basal leaves and shiny, deep yellow petals ¼–⅝" (8–16 mm) long. It is a common introduced weed of roadsides and meadows.

Cursed buttercup (*R. sceleratus*) is 8–24" (20–60 cm) tall and has succulent, deeply lobed leaves, small, ⅛" (2–4 mm) long, pale yellow petals and beakless achenes. It grows in marshy ground from Alaska to New Mexico.

Pasqueflowers

Pulsatilla spp.

All parts of these plants are poisonous if taken internally and irritating if applied externally. The sap contains the glycoside ranunculin (see buttercups, p. 243). Some tribes applied crushed pasqueflower leaves to rheumatic joints, bruises and sore muscles, as a counter-irritant. Homeopaths have used minute doses of prairie crocus to treat eye

Prairie crocus

problems, rashes, rheumatism, menstrual obstruction, bronchitis, asthma and coughs. Some herbalists recommend alcohol extracts from western pasqueflower as an antidepressant sedative. Overdoses cause lowered blood pressure, nausea, salivation and dizziness. This extract should not be used during pregnancy or by people with an abnormally slow heart rate. Sheep have been poisoned by eating prairie crocus in prairie pastures.

DESCRIPTION: Silky-hairy, perennial herbs with finely divided leaves. Flowers with 5 petal-like sepals and no petals, single, above a whorl of stem leaves. Fruits tiny seed-like achenes, each tipped with a long, feathery style, forming fluffy heads when mature.

Prairie crocus or **common pasqueflower** (*P. patens*) is a blue-flowered species of well-drained slopes in plains to subalpine zones from Alaska to New Mexico.

Western pasqueflower (*P. occidentalis*) is a white-flowered species of moist slopes in montane, subalpine and alpine zones in BC, Alberta, Idaho and Montana.

Virgin's-bowers

Clematis spp.

The sap of these plants contains the glycoside ranunculin. Some people develop severe skin reactions and swollen, inflamed eyelids from contact with these plants. If virgin's-bower is eaten, it can cause stomach upset, internal bleeding, nervousness, depression and even death. It has been used to treat nervous disorders, migraine headaches and rashes.

DESCRIPTION: Woody vines with opposites leaves divided into 3–7 stalked leaflets. Flowers with 4 petal-like sepals and no petals. Fruits seed-like achenes, tipped with feathery styles, forming fluffy heads when mature.

White virgin's-bower (*C. ligusticifolia*) has clusters of $^5/_8$–$^3/_4$" (1.5–2 cm) wide, cream-colored flowers. It grows along roads and rivers in the plains and foothills zones from BC and Alberta to New Mexico.

White virgin's-bower

Blue virgin's-bower (*C. occidentalis*) has single, nodding flowers, with purple sepals $1^3/_8$–$2^1/_2$" (3.5–6 cm) long. It grows in open forests in foothills and montane zones from BC and Alberta to Colorado.

Baneberry

Actaea rubra

Baneberry poisoning is usually attributed to an unknown essential oil, but some sources report that these plants contain the glycoside ranunculin (see p. 243). All parts of baneberry are poisonous, but the roots and berries are most toxic. Eating 2–6 berries can cause severe cramps and burning in the stomach, vomiting, bloody diarrhea, increased pulse, headaches and/or dizziness. Severe poisoning results in convulsions, paralysis of the respiratory system and cardiac arrest. No deaths have been reported in North America, probably because the berries are extremely bitter. Native peoples used baneberry-root tea to treat menstrual and postpartum problems, colds, coughs, rheumatism and syphilis. Some herbalists have used baneberry roots as a strong antispasmodic, anti-inflammatory, vasodilator and sedative, usually for treating menstrual cramps and menopausal discomforts.

DESCRIPTION: Branched, leafy, perennial herb, 1–3 1/2" (30–100 cm) tall, with coarsely toothed leaves divided 2–3 times in 3s. Flowers white, with 5–10 slender, 1/16–1/8" (2–3 mm) long petals, forming long-stalked, rounded clusters, in May to July. Fruits glossy red or white berries 1/4–3/8" (6–8 mm) long. Baneberry grows in moist, shady foothill, montane and subalpine sites from the Yukon and NWT to New Mexico.

Columbines

Aquilegia spp.

All columbines are probably somewhat poisonous. The flowers are said to be sweet and edible in small quantities, but consumption is not recommended. All other parts of these plants can be quite toxic. The seeds and roots are most poisonous. In Europe, columbines were used to treat many ailments, including heart palpitations, boils, ulcers, gall or kidney stones and jaundice. However, some children died from overdoses of the seeds and medicinal use was discontinued.

DESCRIPTION: Erect, perennial herbs, 1/2–2 1/4' (20–70 cm) tall. Leaves alternate, divided 2–3 times in 3s, with broad, round-lobed leaflets 3/4–2" (2–5 cm) long. Flowers with 5 spreading, petal-like sepals and 5 smaller petals that extend back from the base as spurs, nodding in small, loose clusters.

Yellow columbine

Colorado columbine (*A. coerulea*) has blue-and-white flowers, 2–4" (5–10 cm) across, with straight spurs twice as long as their petal blades. It grows on moist foothill, montane and subalpine slopes from Idaho and Montana to New Mexico.

Yellow columbine (*A. flavescens*) has yellow flowers, about 2' (5 cm) wide, with incurved spurs. It grows on moist foothill to alpine slopes from BC and Alberta to Colorado.

Mountain monkshood

Monkshoods

Aconitum spp.

All parts of these plants contain the poisonous alkaloid aconitine, plus other toxins. The roots are most poisonous—ingestion of less than 18 oz (500 g) can be fatal to a horse. The flowers are harmless to handle but violently poisonous if eaten. Symptoms include anxiety, vomiting, diarrhea, weakness, salivation, dizziness, numbness and prickling, impaired speech and vision, weak pulse, convulsions, paralysis of the lower legs and respiratory system and coma. Death can occur in a few hours. The sap causes numbness and tingling on the skin.

DESCRIPTION: Erect, perennial herbs with palmately divided leaves. Flowers dark blue to purple, flattened sideways, with a large hood over 2 parallel side wings and 3 small lower petals. Fruits erect groups of pods with spreading tips.

Columbian monkshood (*A. columbianum*) is 1–4¹/₂" (30–130 cm) tall, with deeply lobed leaves (not divided to the very base). It grows on moist foothill, montane and subalpine slopes from BC and Montana to New Mexico.

Mountain monkshood (*A. delphinifolium*) is less than 2' (60 cm) tall, with separate, short-stalked leaflets. It grows on moist subalpine and alpine slopes from Alaska to BC and Alberta.

Tall larkspur

Delphinium spp.

These plants contain many toxic alkaloids. Symptoms of poisoning include burning in the mouth, tingling skin, nausea and cramps, weak pulse, difficulty breathing, nervousness and depression or excitement. Some people develop skin reactions from contact with delphiniums. Toxicity decreases as the plants age, and the seeds are very poisonous. Many cattle have been poisoned by larkspurs. Fatal poisoning usually requires eating plants equivalent to at least 3 percent of an animal's body weight. Sheep tolerate high levels of these toxins, and tall larkspur can provide fair-to-good forage for them. Teas and tinctures made from larkspur seeds were used for many years to kill lice and to cure scabies, but these extracts are now considered too dangerous to use.

DESCRIPTION: Erect perennials with palmately divided leaves. Flowers mostly purplish, with 5 spreading petal-like sepals below 4 small petals. Fruits erect clusters of pods with spreading tips.

Low larkspur (*D. bicolor*) has solid, 4–20" (10–50 cm) tall stems, with small clusters of blue-and-white flowers. It grows on dry plains, montane and subalpine slopes from BC and Alberta to Wyoming.

Tall larkspur (*D. glaucum*) has hollow, 2–6¹/₂' (60–200 cm) tall stems, with larger clusters of purplish-blue flowers. It grows on moist montane, subalpine and alpine slopes from Alaska to Montana.

Lupines

Lupinus spp.

These plants can be especially dangerous because their pods look like hairy garden peas, and children may assume that they are edible. Some lupines are edible, but many contain poisonous alkaloids, and even botanists can have difficulty distinguishing poisonous and non-poisonous species. Poisonous lupines (especially their flowers, pods and seeds) contain the alkaloids lupanine, lupinine and/or sparteine. Large amounts must be eaten in a short time to cause poisoning, but these toxins are not destroyed by drying, so hay with lupine plants can be poisonous to livestock. Silvery lupine and silky lupine have killed sheep, goats, cattle and horses. In Wyoming, silvery lupine may be responsible for poisoning more sheep than any other plant. Both lupines (especially their seeds, pods and young leaves) contain the glycoside in anagyrine in concentrations of 2–3 times higher than those known to cause birth deformities in calves, kids and lambs. Lactating goats pass anagyrine in their milk. At least 1 child was believed to have been born with deformed limbs because her mother drank milk from goats that were grazing on lupines. A few European lupines are grown as a substitute for peas, but our wild lupines are not used in this way. Some edible lupine seeds are sold in health-food stores, but these seeds must be soaked and then boiled in several changes of water to remove bitter toxins, or they can cause dizziness and incoordination. Lupine poisoning may also slow breathing and heart rate.

DESCRIPTION: Large, complex group of mostly perennial herbs with round, alternate leaves divided into lance-shaped, finger-like leaflets. Flowers mostly blue or purple, pea-like, borne in whorls in showy, elongated clusters. Fruits pea-like pods, mostly hairy, about $3/4$–$1 1/4$" (2–3 cm) long in our species.

Silky lupine (*L. sericeus*) has hairs over most of the upper (back) surface of its top petal, whereas **silvery lupine** (*L. argenteus*) is essentially hairless on its top petal and on its upper leaf surfaces. Most leaves of these species are short-stalked on the stem, with 7–9 leaflets. Both species grow on dry plains to montane slopes from BC and Alaska to New Mexico. Silvery lupine occasionally grows in the subalpine zone.

Silky lupine (top); Silvery lupine (above)

247

Prairie goldenbean

Goldenbeans
Thermopsis spp.

Goldenbeans contain toxic alkaloids, as well as anagyrine, which causes calf deformities when eaten by pregnant cows. Mountain goldenbean has poisoned cattle and horses, and as little as 18 oz (500 g) of ripe seeds can be lethal. The toxicity of prairie goldenbean is less studied. Some sources report that prairie goldenbean is only suspected of poisoning children, but others say that in some regions it causes more human poisonings than any other plant, and that large doses can be fatal.

DESCRIPTION: Perennial herbs with bright yellow, pea-like flowers. Leaves alternate, with 3 oval leaflets and 2 large, leaflet-like basal lobes (stipules). Fruits pea-like pods.

Mountain goldenbean (*T. montana*) is 1¹/₂–3' (50–90 cm) tall, with straight, erect pods. It grows on open foothill and montane slopes from BC to Colorado.

Prairie goldenbean or **buffalobean** (*T. rhombifolia*) is 8–16" (20–40 cm) tall, with hanging, curved pods. It grows in plains to subalpine zones from BC and Alberta to Colorado.

Showy locoweed

Locoweeds
Oxytropis spp.

Many locoweeds are poisonous to horses, sheep and cattle. These plants contain toxic alkaloids, and some species also take selenium from the soil. After eating large quantities of locoweed for several weeks, cattle, horses and sheep develop locoism, a disease that mainly affects the nervous system. Some animals grow to like these plants, and many have died as a result. Symptoms include depression, incoordination and excitability. Locoweeds also cause heart disease, fluid retention, diarrhea, birth deformities and miscarriages. Cattle usually die after eating an equivalent of 300 percent of their body weight, but eating only 30 percent of body weight can be fatal to horses.

DESCRIPTION: Tufted perennial herbs with basal, pinnately divided leaves. Flowers about ³/₈–³/₄" (1–2 cm) long, pea-like, the lower 2 petals forming a pointed keel. Fruits pea-like pods, about ³/₈–³/₄" (1–2 cm) long.

Silky locoweed (*O. sericea*) has yellow flowers, about ³/₄" (2 cm) long, and bony mature pods. It grows on dry foothill, montane and subalpine slopes from the Yukon and NWT to New Mexico.

Lambert's locoweed (*O. lambertii*) has bright rose-purple flowers and flat-lying, parallel hairs. It grows on dry plains and foothill slopes from Montana to Colorado.

Showy locoweed (*O. splendens*) has grayish, silky-hairy plants, with pinkish-purple flowers and whorled leaflets. It grows on well-drained foothill and montane slopes from Alaska to New Mexico.

Timber milk-vetch
Astragalus miser

The milk-vetches comprise a large, complex group of plants, including both harmless and extremely poisonous species. Some become toxic by absorbing harmful chemicals from the soil—selenium causes depression, diarrhea, increased urination, hair loss and death from lung and heart failure; molybdenum causes poor growth, brittle bones and anemia. Some milk-vetches contain the alkaloid locoine, which causes locoism (see locoweed, p. 248), and consequently milk-vetches are often called 'locoweeds.' Timber milk-vetch contains miserotoxin, which can cause acute poisoning (with death a few hours after ingestion) or long-term poisoning (with liver and nerve damage, local bleeding in the brain and difficulty breathing). Honeybees have been poisoned by milk-vetch nectar.

DESCRIPTION: Perennial herb with leafy stems. Leaves pinnately divided into 7–12 leaflets. Flowers white to lilac with blue lines, pea-like, $^3/_8$–$^1/_2$" (8–12 mm) long. Fruits thin, hanging, stalkless pods. It grows on open foothill and montane slopes from BC and Alberta to Colorado.

Peavines
Lathyrus spp.

Many plants of this genus have been used for food (as greens and as peas) and for forage. However, peavines are generally viewed with suspicion, because the seeds of some species contain peculiar amino acids that poison nerve cells (neurotoxins). Eaten in moderation, peavines can provide a nutritious food, but if the diet consists almost exclusively of peavine for 10 days to 4 weeks, a type of poisoning called 'lathyrism' can result. Even the common garden pea (*Pisum* spp.) can cause nervous disorders if large amounts are eaten regularly for long periods of time. Lathyrism causes a progressive loss of coordination, ending in irreversible paralysis. During famines people were forced to eat peavines almost exclusively.

Creamy peavine

DESCRIPTION: Climbing, perennial herbs with loosely clustered, pea-like flowers about $^1/_2$" (10–15 mm) long. Leaves pinnately divided, tipped with tendrils. Fruits slender pods, about $1^1/_2$–$2^1/_2$" (4–6 cm) long.

Creamy peavine or **vetchling** (*L. ochroleucus*) has yellowish-white flowers, ovate leaflets and well-developed tendrils. It grows in moist plains and foothill sites from the Yukon and NWT to Montana.

White-flowered peavine or **vetchling** (*L. lanszwertii*) has linear-oblong leaflets and simple to almost bristle-like tendrils. It grows on well-drained foothill and montane slopes from Wyoming to New Mexico.

American vetch
Vicia americana

The toxicity of vetches varies from person to person and species to species. Some vetches contain the toxic alkaloids vicine and convicine. These alkaloids produce oxidants that attack the red blood cells of people who lack the enzyme glucose-6-phosphate dehydrogenase. Eating the beans or inhaling the pollen of toxic species causes headaches, weakness, dizziness, vomiting, fever, jaundice, anemia and even death in susceptible people. This enzyme deficiency is genetically controlled and is most common in a small percentage of people of Mediterranean or African origin. Domestic broad beans (*V. faba*) cause this reaction (favism), and large quantities of broad beans have poisoned pigs and chickens. Some vetches contain chemicals that produce hydrocyanic acid, and some contain toxic amino acids that affect the nervous system. Vetch seeds are attractive to young children, because they resemble small peas. Young shoots and tender seeds of American vetch have been boiled or baked for food, but these plants should be used with caution, if at all, considering the history of their relatives.

DESCRIPTION: Slender, vine-like herb with tangled, 4-sided stems. Leaves pinnately divided, tipped with tendrils. Flowers reddish-purple, pea-like, 5/8–3/4" (15–20 mm) long, in loose clusters of 2–9. Fruits flat, hairless pods 3/4–1 1/4" (2–3 cm) long. American vetch grows on moist plains, foothill and montane sites from the NWT to New Mexico.

Leafy spurge
Euphorbia esula

The acrid milky juice (latex) of leafy spurge can inflame and blister sensitive skin. Most animals (including humans) do not eat these plants. Symptoms of poisoning include burning in the mouth and throat, swelling around the mouth and nose, abdominal pains, vomiting, diarrhea and fainting spells. Even honey made from the nectar of some spurges is mildly poisonous. Sheep are relatively resistant to the toxins, so they have been used, in conjunction with beetles, in biological control programs. However, large quantities can kill sheep. Spurge toxins are not neutralized by drying, so these plants remain toxic in hay. Non-fatal internal doses can cause skin reactions in people, sheep, cattle and horses, after exposure to sunlight. Deep rootstocks and prolific seed production make leafy spurge difficult to control, and it is considered a noxious weed in many regions.

DESCRIPTION: Erect, bluish-green, perennial herb with milky juice. Leaves linear, 3/4–2 1/2" (2–6 cm) long. Flowers yellowish-green, tiny, but borne above showy pairs of heart-shaped, yellowish-green bracts that resemble petals. Fruits round, finely granular capsules about 1/8" (4 mm) wide. Leafy spurge was introduced from Europe and now grows on disturbed sites in the plains and foothills zones from BC and Alberta to New Mexico.

European bittersweet

Solanum dulcamara

The immature (green) berries and leaves contain toxic alkaloids and plant sterols, which can cause vomiting, dizziness, weakened heart, liver damage, convulsions, paralysis and death. Cattle and sheep have been poisoned by eating these bitter plants, but deaths are rare. The bright red, ripe berries contain only small amounts of alkaloids and are not considered a threat if eaten in moderation, but large amounts could prove toxic. Stem extracts have been taken internally as a sedative and pain reliever, for increasing urination and for treating asthma, gout, rheumatism, whooping cough and bronchitis, but pharmacological evidence does not support these uses. Extracts are reported to have antibiotic activity, which could be useful in salves and lotions for combating infection. These plants have been used for many years to treat skin diseases, sores, swellings and inflammations around nails. Recent research has shown that bittersweet contains beta-solanine, a tumor-inhibiting compound that may prove useful for treating cancer.

DESCRIPTION: Scrambling, semi-woody, perennial herb with oval to 3-lobed leaves 1^1/$_4$–3" (3–8 cm) long. Flowers resemble small shooting stars, with a yellow cone projecting from the center of 5 back-curved, blue-violet petals, borne in loose, branched clusters. Fruits bright red, oblong berries. This introduced weed grows on moist, disturbed ground at low elevations from BC to New Mexico.

Arnicas

Arnica spp.

All arnicas can cause severe upset and blistering of the digestive tract. Flower and rootstock extracts are said to dilate capillaries under the skin and to stimulate the activity of white blood cells. Arnicas have been used for centuries in liniments, salves, washes and poultices for treating bruises, chilblains, sprains and swollen feet and for stimulating hair growth. These plants should never be applied to broken skin, where toxins could enter the bloodstream. A few people develop severe skin reactions from contact with arnicas.

DESCRIPTION: Perennial herbs with opposite leaves and yellow, sunflower-like flowerheads. Fruits seed-like achenes with fluffy white or tawny parachutes.

Leafy arnica (*A. chamissonis*) is 8–40" (20–100 cm) tall and has 5–10 pairs of leaves below branched clusters of flowerheads. Each involucral bracts is tipped with white hairs. It grows in moist foothill to subalpine sites from the Yukon and NWT to New Mexico.

Heart-leaved arnica (*A. cordifolia*) is 4–24" (10–60 cm) and has long-stalked, heart-shaped basal leaves. It grows on foothill to subalpine slopes from the Yukon to New Mexico.

251

Groundsels

Senecio spp.

Some groundsels (e.g., common groundsel, western groundsel) contain pyrrolizidine alkaloids that cause irreversible liver damage after long-term exposure and may cause liver cancer. Alkaloid concentrations are highest in the flowers and lowest in roots and young leaves, but levels vary greatly from species to species and plant to plant. These toxins are not destroyed by drying. Horses and cattle have died from eating common groundsel. Young animals are most susceptible. Unfortunately, liver damage is usually severe before outward signs are noticeable. Some people use groundsel in teas and herbal remedies. Extended use of groundsel tea or use of flour contaminated with groundsel seed has caused loss of appetite, vomiting, bloody diarrhea, sleepiness, weakness, staggering and jaundice, with liver damage and even death. Some people develop rashes from contact with these plants. Pregnant women should not use groundsel tea because it could harm the fetus. The bitter-tasting honey made from groundsel nectar contains high levels of

Common groundsel (top); Arrow-leaved groundsel (above)

pyrrolizidine alkaloids. These toxins are also passed in the milk of dairy animals. Very few of the wild groundsels in the Rocky Mountains are known to have high alkaloid levels, but not all have been analyzed. Some species (e.g., arrow-leaved groundsel) are recommended as wild greens, but without further information it is probably best to view all members of this genus with suspicion.

DESCRIPTION: One of the largest and most diverse genera in the world. Species in the Rockies are annual or perennial herbs with alternate leaves, few to many small, yellow flowerheads, involucral bracts in a single row and seed-like achenes with fluffy, white 'parachutes.'

Common groundsel (*S. vulgaris*) is an annual species with inconspicuous flowerheads that lack ray florets. This introduced weed grows on disturbed ground from BC and Alberta to New Mexico.

Western groundsel (*S. integerrimus*) has toothless leaves and black-tipped involucral bracts. It grows in plains to subalpine zones from BC and Alberta to Colorado.

Arrow-leaved groundsel (*S. triangularis*) has sharply toothed, triangular leaves. It grows on moist foothill to alpine sites from the Yukon and NWT to New Mexico.

GLOSSARY

achene: a small, dry fruit that doesn't split open, often seed-like in appearance, distinguished from a nutlet by its relatively thin wall.

alkaloid: any of a group of bitter-tasting, usually mildly alkaline plant chemicals. Many alkaloids effect the nervous system.

allergen: a substance that causes an allergic reaction.

alterative: a medicine that positively changes the course of an ailment.

alternate: situated singly at each node or joint (e.g., as leaves on a stem) or regularly between other organs (e.g., as stamens alternate with petals).

anaphylaxis: increased sensitivity to a foreign substance (often a protein) resulting from previous exposure to it (as in serum treatment).

andromedotoxin: a toxic alkaloid derived from a diterpene.

annual: a plant that completes its life cycle in 1 growing season.

anti-coagulant: an agent that stops or slows coagulation.

antihelminthic: an agent that expels worms.

antimocrobial: an agent that destroys or prevents the growth of microorganisms.

arbutin: a glycoside found in plants of the Heath family (Ericaceae).

aril: a specialized covering attached to a mature seed.

astringent: an agent that contracts body tissues and checks secretions, capillary bleeding, etc.

axil: the position between a side organ (e.g., a leaf) and the part to which it is attached (e.g., a stem).

balm: a healing or soothing ointment.

bannock: a flat bread, traditionally unleavened but now made with flour, fat and baking powder.

barbiturate: an organic derivative of barbituric acid that depresses the central nervous system, respiration, heart rate, blood temperature and blood pressure.

beak: a prolonged, more or less slender tip on a thicker organ such as a fruit or seed.

biennial: living for 2 years, usually producing flowers and seed in the second year.

boil: painful, localized inflammation of the lower layers of a skin gland or hair follicle, producing a hard central core and pus, usually caused by a bacterial (staphylococcus) infection.

bolt: to send up long, erect stems (usually elongating flower/fruit clusters) from a basal rosette.

bract: a specialized leaf with a flower (or sometimes a flower cluster) arising from its axil.

bulb: a short, vertical underground stem with thickened leaves or leaf bases (e.g., an onion).

bulblet: a small bulb-like structure produced in a leaf axil or replacing a flower.

burl: a hard, often round, woody outgrowth on a tree.

bursitis: inflammation of a pad-like sac (bursa) in connecting tissue, usually around joints.

calyx: the outer (lowermost) circle of floral parts, composed of separate or fused lobes called sepals, usually green and leaf-like.

cambium: the thin growing layer, responsible for producing new stem cells.

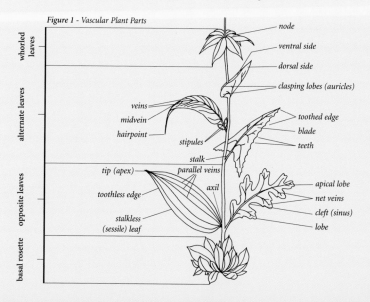

Figure 1 - Vascular Plant Parts

whorled leaves — node, ventral side, dorsal side, clasping lobes (auricles)

alternate leaves — veins, midvein, hairpoint, stipules, stalk, toothed edge, blade, teeth

opposite leaves — tip (apex), parallel veins, toothless edge, axil, stalkless (sessile) leaf, apical lobe, net veins, cleft (sinus), lobe

basal rosette

candidiasis: an infection of skin or mucous membranes caused by a species of the yeast-like fungus, Candida.

carbuncle: a painful localized inflammation of skin and deeper tissues with several openings for the discharge of pus and with sloughing of dead tissue.

cardiac; cardio-: of the heart.

catkin: a dense spike or raceme of many small, unisexual, naked flowers that lack petals and sepals but have a bract.

cellulose: the polysaccharide that comprises most of plant cell walls.

chiggers: the 6-legged, parasitic larvae of mites in the family Trombiculidae; also called redbugs.

chlorophyll: the pigment that gives most plants their green color and the means of manufacturing sugars and starches through photosynthesis.

chronic: persistent, of long duration.

cirrhosis: loss of function of an organ; usually applied to progressive liver dysfunction because of development of dense nodules and fibers.

clasping: embracing or surrounding, usually in reference to a leaf base around a stem.

corm: a swollen stem base containing food material and bearing buds in the axils of the scale-like remains of leaves from the previous season.

corolla: the second circle of floral parts, composed of separate or fused lobes called petals; usually conspicuous in size and color but sometimes small, reduced to nectaries or absent.

coumarin: a white, crystalline lactone with the sweet smell of new-mown hay; a toxic anti-coagulant.

counter-irritant: an agent for producing irritation in 1 part to counteract irritation or relieve pain or inflammation elsewhere.

cultivar: a plant or animal originating in cultivation.

cyst: a closed sac containing fluid, semi-fluid or solid material, usually an abnormal growth caused by developmental anomalies, duct obstruction or parasitic infection.

decoction: a solution produced by boiling a plant substance in water.

delirium tremens: a psychic disorder, usually associated with withdrawal from alcohol, producing visual and auditory hallucinations.

dermatitis: inflammation of the skin, with redness, itching and/or lesions.

digitalis: the powdered leaves of common foxglove (*Digitalis purpurea*), containing glycosides that act as a powerful heart stimulant.

disc floret: a small, tubular flower in a flowerhead of the Aster family, usually clustered at the center of the head.

diuretic: an agent that increases the flow of urine.

Doctrine of Signatures: a medical theory first popularized in the 14th century, in which each plant displays a clear sign or 'signature' of the purpose for which it was intended: plants with heart-shaped leaves being good for the heart; those resembling hair being good for the hair, etc.

dropsy: an obsolete term or edema.

drupe: a fruit with an outer fleshy part covered by a thin skin and surrounding a hard or bony stone that encloses a single seed (e.g., a plum).

druplet: a tiny drupe, part of an aggregate fruit such as a raspberry.

dysentery: intestinal disorders, especially of the colon, caused by infection and characterized by severe bloody diarrhea.

eczema: a disease of the skin, characterized by inflammation, itching and the formation of scales.

edema: the abnormal accumulation of excessive amounts of fluids in body tissues.

emmenagogue: an agent that assists or promotes menstrual flow.

Figure 2 - Underground Parts

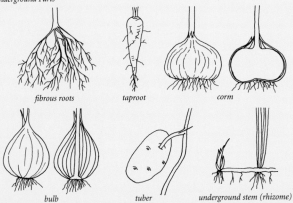

fibrous roots   taproot   corm

bulb   tuber   underground stem (rhizome)

emphysema: a condition (usually of the lungs) characterized by air-filled expansions within tissues that may damage the tissue structure.

endocrine: pertaining to a gland that secretes directly into the blood stream.

enema: injection of liquid into the rectum and colon, via the anus.

ergot: a fungal disease of grasses caused by *Claviceps purpurea*, with black to dark purple fruiting bodies containing several powerful alkaloids.

excelsior: fine, curled wood shavings, often used for packing fragile items.

expectorant: an agent that facilitates removal (coughing-up) of mucus from the respiratory tract.

febrifuge: an agent that reduces fever.

flavone: a colorless, crystalline ketone found in many primroses; flavone derivatives often occur in yellow pigments and are used in dyes.

floret: a small flower, usually 1 of several in a cluster.

follicle: a dry, pod-like fruit, splitting open along a single line on 1 side.

forb: a broad-leaved flowering plant, as distinguished from grasses, sedges, etc.

frond: a fern leaf.

fructose: a 5-carbon monosaccharide found in corn syrup, honey, fruit juices, etc; fruit sugar.

fumigant: smoke, vapor or gas applied to kill pests or combat infection.

gingivitis: gum disease producing redness, swelling and a tendency to bleed.

glume: one of a pair of bracts at the base of a grass spikelet, often persisting after the seeds have fallen.

glycoside: a 2-parted molecule composed of a sugar and an aglycone, usually becoming poisonous when digested and the sugar is separated from its poisonous aglycone.

gout: a hereditary metabolic disorder, a form of acute arthritis, with painful inflammation of joints (usually in the knees or feet).

gram-positive: retaining the purple color when stained with gentian violet in the Gram's test, usually pertaining to bacteria.

gravel: crystalline dust or stones in the kidney, composed of phosphate, calcium, oxalate and uric acid.

gruel: thin porridge.

haw: the fruit of a hawthorn, usually with a fleshy outer layer enclosing many dry seeds.

hemorrhoid: a mass of dilated, twisted veins in the lower rectum–upper anus.

histamine: a compound that dilates capillaries and contracts smooth muscles, often released during allergic reactions.

hives: itchy weals caused by an allergic reaction.

hybrid: a cross between 2 species.

hyperglycemia: too much sugar in the blood.

hyperthyroidism: excessive secretion by the thyroid glands, resulting in an increased metabolic rate.

hypoglycemia: too little sugar in the blood.

incontinent: unable to retain bodily discharges, such as urine or feces.

inflammation: redness, pain, heat, swelling and/or loss of function of some part of the body, as a result of injury, infection, irritation, etc.

inflated: distended with air or gas, blown up like a balloon.

infusion: a liquid extract made by steeping a substance in water that was initially boiling.

involucre: a set of bracts closely associated with one another, encircling and immediately below a flower cluster.

involuntary muscle: a muscle controlling a reflex action and not under direct, voluntary control, especially a smooth muscle.

irregular: usually referring to bilaterally symmetrical flowers in which the upper half is unlike the lower, while the left half is a mirror image of the right; occasionally regarding flowers in which members of 1 or more sets of organs differ among themselves in size, shape or structure.

Figure 3 - Parts of an Aster (Asteraceae) Flowerhead

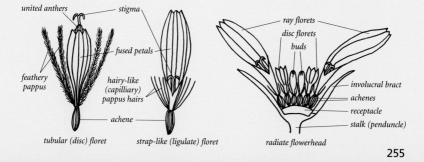

united anthers — stigma

fused petals

feathery pappus

hairy-like (capilliary) pappus hairs

achene

tubular (disc) floret   strap-like (ligulate) floret

ray florets

disc florets

buds

involucral bract

achenes

receptacle

stalk (penduncle)

radiate flowerhead

255

iso-: of equal or homogeneous form.

jaundice: an abnormal condition caused by increased bile pigments in the blood, characterized by yellowish skin, yellowish whites of the eyes, weariness and loss of appetite.

keel: a sharp, conspicuous longitudinal ridge, like the keel of a boat; the 2 partly united lower petals in a pea flower.

lactation: the secretion of milk.

latex: milky plant juice containing rubbery compounds.

lecithin: a fatty substance (phospholipid) found in blood, bile, nerves and other animal tissues.

lemma: the lower of the 2 bracts immediately enclosing a grass flower.

lenticel: a slightly raised pore on root, trunk or branch bark.

lignin: the substance that bonds with cellulose to form woody cell walls and to cement these cells together.

liniment: a liquid medication applied externally by rubbing or by applying on a bandage.

linoleic acid: an unsaturated fatty acid, essential for the nutrition of some animals.

lip: a projection or expansion of something, such as the lower petal of a violet flower.

lipid: any of a group of fat-like substances (including true fats, lipoids and sterols) that are insoluble in water and soluble in fat solvents, such as alcohol.

lupus: any of a group of diseases characterized by skin lesions.

melanoma: a malignant, darkly pigmented tumor or mole.

meprobamate: a bitter tranquilizer used to relieve anxiety.

mordant: a substance used to fix color when dyeing, often a metallic agent that combines with a dye to form an insoluble, colored compound.

mucilage: a sticky, gelatinous plant substance.

mutagen: an agent that causes genetic mutations.

narcotic: producing stupor or sleep and reducing pain.

neuralgia: severe pain along the course of a nerve or in its area of distribution.

neurotoxin: a nerve poison.

nutlet: a small, hard, dry, 1-seeded fruit or part of a fruit, not splitting open.

opposite: situated across from each other at the same node (not alternate or whorled); or situated directly in front of another organ (e.g., stamens opposite petals).

oxalate: a salt or ester of an oxalic acid.

oxalic acid: a strong, poisonous acid, often used in white, crystalline powder form as a bleach or stain remover.

oxytocin: a pituitary hormone that stimulates contraction of the uterus and secretion of milk.

palea: the upper of the 2 bracts immediately enclosing a grass flower.

palmate: divided into 3 or more lobes or leaflets diverging from a common point, like fingers on a hand.

palpitation: strong, rapid pulsation, such as a fast throbbing or fluttering of the heart.

panicle: a loosely branched cluster of stalked flowers or spikelets, blooming from the bottom up.

pappus: a modified calyx forming a crown of bristles, hairs, awns, teeth or scales at the tip of the seed, often aiding in seed dispersal.

parch: to toast under dry heat.

pectin: a water-soluble substance that binds adjacent cell walls together, producing a gel.

pemmican: a mixture of finely pounded, dried meat, fat and sometimes dried fruit.

perennial: living for 3 or more years, usually flowering and fruiting for several years.

Figure 4 - Parts of a Female Willow (Salix) *Shrub*

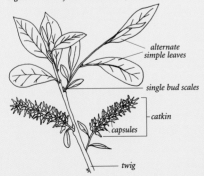

alternate simple leaves

single bud scales

catkin

capsules

twig

Figure 5 - Branching Patterns

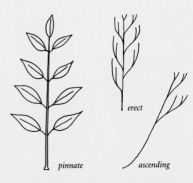

erect

pinnate

ascending

256

perigynium: a modified bract forming a sac around the pistil (achene) of a sedge (*Carex*) flower.

pharmacology: the science of drugs, their properties and reactions, especially those related to therapeutic use.

photosynthesis: the process by which plants manufacture sugars and starches using chlorophyll, carbon dioxide, water and the energy of the sun.

pinnate: with branches, lobes, leaflets or veins arranged on both sides of a central stalk or vein; feather-like.

pinole: flour made from finely ground, parched corn.

plaque: a localized, abnormal patch on skin or mucous membranes, often referring to abnormal fatty deposits in arteries.

platelets: minute discs in the blood that play an important role in blood coagulation.

pome: a fleshy fruit with a core (e.g., an apple), comprised of an enlarged hypanthium around a compound ovary.

postpartum: after childbirth.

poultice: a soft, moist mass of cloth, bread, meal or herbs, applied as a remedy for sore or inflamed parts of the body.

prostaglandin: any of a group of compounds derived from saturated fatty acids and found in many body tissues, having hormone-like actions, including control of uterine contractions and blood pressure.

psoralen: any of a group of plant substances, some of which are phototoxic (causing skin reactions when applied to skin and then exposed to ultraviolet radiation).

psoriasis: a chronic skin disease characterized by scaly, reddish patches.

purgative: causing watery evacuation of the bowels.

quinine: a bitter, white crystalline alkaloid derived from cinchona bark, used to treat malaria.

raceme: an unbranched cluster of stalked flowers on a common, elongated central stalk, blooming from the bottom up.

ray floret: a small, flattened, strap-like flower of a flowerhead in the Aster family, often radiating from the edges of the head.

receptacle: an expanded stalk tip at the center of a flower, bearing the floral organs or the small, crowded flowers of a head.

rickets: a deficiency disease in which lack of sun light and vitamin D results in insufficient assimilation of calcium and phosphorus into newly formed bones, causing abnormalities in bone shape and structure.

ringworm: an itchy, red-ringed, painful, scaly fungal infection of the skin, including the scalp (tinea capitis), body (tinea corporis), beard (tinea barbae), nails (tinea unguium) and feet (athlete's foot).

rosette: a cluster of crowded, usually basal leaves in a circular arrangement.

salicin: a glycoside found in willows and poplars, related to salicylic acid and used, like aspirin, for relieving pain and fever.

saponin: any of a group of glycosides with steroid-like structure found in many plants, causing diarrhea and vomiting when taken internally, often used as detergents.

scabies: a contagious skin disease caused by a parasitic mite (*Sarcoptes scabiei*) that burrows under the skin to deposit eggs, causing intense itching.

schizocarp: a fruit that splits into 2 or more parts at maturity (e.g., fruits of the Carrot family [Apiaceae]).

sciatica: a painful condition of the back of the hips and thigh.

scrofula: tuberculosis of the lymph nodes, especially in the neck.

scurvy: a deficiency disease caused by lack of vitamin C, characterized by bleeding and abnormal formation of bones and teeth.

Figure 6 - Fruits

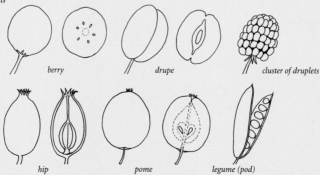

berry   drupe   cluster of druplets

hip   pome   legume (pod)

sepal: 1 segment of the calyx, usually green and leaf-like.

shingles: a disease caused by reactivation of the chickenpox virus (*Herpes zoster*), characterized by painful, inflamed eruptions along a peripheral nerve, usually on the trunk of the body (occasionally on the head).

sitz bath: a tub for sitting in, with water covering the hips.

smudge: a smoldering mass, placed on the windward side to produce smoke and/or heat for protection from insects, frost, etc.

snoose: a small wad of tobacco held between the gums and the lip or cheek.

spike: a simple, unbranched flower cluster, with (essentially) stalkless flowers arranged on an elongated axis.

spikelet: a small or secondary spike, as in the flowerheads of grasses and sedges.

sporangium: a structure in which spores are produced; a spore-case.

spur: a hollow appendage on a petal or sepal, usually functioning as a nectary.

steroid: any of many compounds containing a 17-carbon, 4-ring structure, including D vitamins, saponins, glucosides of digitalis and certain hormones and carcinogens.

sterol: a solid steroidal alcohol (such as cholesterol) found in plant and animal fats.

stipules: a pair of bract-like or leaf-like appendages at the base of a leaf stalk.

stolon: a slender, prostrate, spreading branch, rooting and often developing new shoots and/or plants at its nodes or at the tip.

style: the part of the pistil connecting the stigma to the ovary, often elongated and stalk-like.

tallow: the solid, nearly tasteless fat rendered from animal fat.

tannin: a soluble, astringent phenol found in many plants, used in tanning, dyeing and medicine.

taproot: a root system with a prominent main root, directed vertically downwards and bearing smaller side roots, sometimes becoming very swollen and containing stored food material (e.g., starch or sugar).

tartar: an encrustation on the teeth, created by food, saliva and salts, such as calcium carbonate.

tendinitis: inflammation of a tendon.

tendril: a slender, clasping or twining outgrowth from a stem or leaf.

tepal: a sepal or petal, when these structures are not easily distinguished.

throat: the opening into a corolla tube or calyx tube.

tincture: an alcohol or alcohol-and-water solution containing animal, plant or synthetic drugs.

tooth: a small, often pointed lobe on the edge of a plant organ (usually on a leaf).

tuber: a thickened portion of a below-ground stem or root, serving for food storage and often also for propagation.

tubercle: a small swelling or projection on an organ.

ulcer: an inflamed lesion of the skin or of a mucous membrane, with sloughing dead tissue and often pus.

umbel: a round- or flat-topped flower cluster in which all flower stalks are of similar length and arise from the same point.

vasoconstrictor: an agent causing constriction of blood vessels.

vasodilator: an agent causing dilation of blood vessels.

vein: a strand of conducting tubes (a vascular bundle), especially if visible externally, as in a leaf.

whorl: a ring of 3 or more similar structures (e.g., leaves, branches or flowers) arising from 1 node.

wing: a thin, flattened expansion on the side(s) or tip of an organ; the side petals of most flowers of the Pea family (Fabaceae).

winnow: to remove lighter material, such as chaff, using wind currents.

Figure 6 (cont.) - Fruits

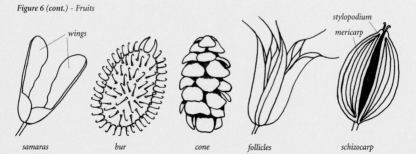

samaras   bur   cone   follicles   schizocarp

REFERENCES

Alberta Agriculture. 1983. *Outdoor Plants Harmful or Poisonous to Humans.* Agdex 666-2. Alberta Tree Nursery and Horticulture Centre, Alberta Agriculture, Edmonton.

Blackwell, Will H. 1990. *Poisonous and Medicinal Plants.* Prentice Hall, Engelwood Cliffs, New Jersey.

Bush, C. Dana. 1990. *A Compact Guide to Wildflowers of the Rockies.* Lone Pine Publishing, Edmonton.

Coffey, Timothy. 1993. *The History and Folklore of North American Wildflowers.* Facts On File, New York.

Craighead, J. J., F. C. Craighead, Jr., and R. J. Davis. 1963. *A Field Guide to Rocky Mountain Wildflowers.* Houghton Mifflin Company, Boston.

Crellin, John K., and Jane Philpot. 1990. *Herbal Medicine Past and Present. Volume 2: A Reference Guide to Medicinal Plants.* Duke University Press, Durham, North Carolina.

Dennis, La Rea L. 1980. *Name Your Poison: A Guide to Cultivated and Native Oregon Plants Toxic to Humans.* Oregon State University Book Stores, Corvallis.

Densmore, F. 1928. *How Indians Use Wild Plants for Food, Medicine and Crafts.* Dover Publications, New York.

Duke, James A. 1986. *Handbook of Northeastern Indian Medicinal Plants.* Quarterman Publications, Lincoln, Massachusetts.

———. 1992. *Handbook of Edible Weeds.* CRC Press, Ann Arbor, Michigan.

———. 1997. *The Green Pharmacy.* Rodale Press, Emmaus, Pennsylvania.

Elias, Thomas S., and P. A. Dykeman. 1990. *Edible Wild Plants of North America Field Guide.* An Outdoor Life Book. Sterling Publishing Company, New York.

Elmore, F. H. 1944. *Ethnobotany of the Navajo.* University of New Mexico Bulletin: Monograph Series, Vol. 1, No. 7. University of New Mexico Press, Albuquerque.

Erichsen-Brown, Charlotte. 1979. *Medicinal and Other Uses of North American Plants.* Dover Publications, New York.

Fernald, M. L., and A. C. Kinsey. *Edible Wild Plants of Eastern North America.* Revised by R. C. Rollins. 1958. Harper and Brothers Publishers, New York.

Foster, S., and James A. Duke. 1990. *Eastern/Central Medicinal Plants.* Peterson Field Guide Series. Houghton Mifflin Company, Boston.

Grieve, M. 1931. *A Modern Herbal.* Penguin Books, Harmondsworth, Middlesex, England.

Hall, Alan. 1973. *The Wild Food Trail Guide.* Holt, Rinehart and Winston, New York.

Hardin, J. W., and J. M. Arena. 1969. *Human Poisoning from Native and Cultivated Plants.* Duke University Press, Durham, North Carolina.

Harrington, H. D. 1967. *Edible Native Plants of the Rocky Mountains.* University of New Mexico Press, Albuquerque.

Hart, Jeff. 1992. *Montana: Native Plants and Early Peoples.* Montana Historical Society Press, Helena, Missouri.

Johnson, Derek, Linda Kershaw, Andy MacKinnon and Jim Pojar. 1995. *Plants of the Western Boreal Forest and Aspen Parkland.* Lone Pine Publishing, Edmonton.

Johnston, A. 1987. *Plants and the Blackfoot.* Occasional Paper No. 15. Lethbridge Historical Society, Lethbridge, Alberta.

Kerik, J., and S. Fisher. 1982. *Living With the Land: Use of Plants by the Native People of Alberta.* Provincial Museum of Alberta, Edmonton.

Kershaw, Linda. 1991. *The Plants of Northwestern Canada, with Special Reference to the Dempster Highway, Yukon and Northwest Territories.* Unpublished manuscript.

Kershaw, Linda, Andy MacKinnon and Jim Pojar. 1998. *Plants of the Rocky Mountains.* Lone Pine Publishing, Edmonton.

Kindschner, K. 1987. *Edible Wild Plants of the Prairie: An Ethnobotanical Guide.* University Press of Kansas.

Kirk, Donald R. 1975. *Wild Edible Plants of Western North America.* Naturegraph Publishers, Happy Camp, California.

Kunkel, G. 1984. *Plants for Human Consumption: An Annotated Checklist of the Edible Phanerogans and Ferns.* Koeltz Scientific Books, Koenigstein, Federal Republic of Germany.

Langshaw, R. 1983. *Naturally: Medicinal Herbs and Edible Plants of the Canadian Rockies.* Summer-thought Publications, Banff, Alberta.

Lyle, Katie L. 1994. *The Wild Berry Book: Romance, Recipes and Remedies.* NorthWord Press, Minocqua, Wisconsin.

MacKinnon, Andy, Jim Pojar and Ray Coupé. 1992. *Plants of Northern British Columbia.* Lone Pine Publishing, Edmonton.

Medsger, O. P. 1939. *Edible Wild Plants.* Collier-Macmillan Canada, Toronto.

Meunscher, Walter Conrad. 1943. *Poisonous Plants of the United States.* The Macmillan Company, New York.

Moore, Michael. 1979. *Medicinal Plants of the Mountain West.* Museum of New Mexico Press, Santa Fe.

———. 1993. *Medicinal Plants of the Pacific West.* Red Crane Books, Santa Fe.

Mulligan, G. A., and D. B. Munro. 1990. *Poisonous Plants of Canada.* Publication 1842/E. Agriculture Canada, Ottawa.

REFERENCES

Munro, Derek B. 1993. *Canadian Poisonous Plants Information System*. Information Systems 1993-1B. Minister of Supply and Services Canada, Ottawa.

Naegele, Thomas A. 1996. *Edible and Medicinal Plants of the Great Lakes Region*. Wilderness Adventure Books, Davisburg, Michigan.

Peirce, A. 1999. *The American Pharmaceutical Association Practical Guide to Natural Medicines*. The Stonesong Press, William Morrow and Company, New York.

Peterson, L. A. 1977. *A Field Guide to Edible Wild Plants of Eastern and Central North America*. Peterson Field Guide Series. Hougton Mifflin Company, Boston.

Pojar, Jim, and Andy MacKinnon. 1994. *Plants of Coastal British Columbia, including Washington, Oregon and Alaska*. Lone Pine Publishing, Edmonton.

Pond, Barbara. 1974. *A Sampler of Wayside Herbs*. The Chatham Press, Riverside, Connecticut.

Readers' Digest. 1986. *Magic and Medicine of Plants*. Readers' Digest Association, Pleasantville, New York.

Robbins, W. W., J. P. Harrington and B. Freire-Marreco. 1916. *Ethnobotany of the Tewa Indians*. Bulletin 55. Bureau of American Ethnology, Smithsonian Institution, Washington, D.C.

Robinson, Peggy. 1979. *Profiles of Northwest Plants: Food Uses, Medicinal Uses, Legends*. Far West Book Service, Portland, Oregon.

Rogers, Dilwyn J. 1980. *Edible, Medicinal, Useful and Poisonous Wild Plants of the Northern Great Plains–South Dakota Region*. Biology Department, Augustana College, Sioux Falls, South Dakota.

Stephens, H. A. 1980. *Poisonous Plants of the Central United States*. The Regents Press of Kansas, Lawrence.

Tanaka, T. 1976. *Tanaka's Cyclopedia of Edible Plants of the World*. Keigaku Publishing Company, Tokyo.

Taylor, Ronald J. 1990. *Northwest Weeds: The Ugly and Beautiful Villains of Fields, Gardens and Roadsides*. Mountain Press Publishing Company, Missoula, Montana.

Thieret, J. W. 1956. Bryophytes as Economic Plants. *Economic Botany* 10: 75–91.

Tilford, Gregory L. 1997. *Edible and Medicinal Plants of the West*. Mountain Press Publishing Company, Missoula, Montana.

Turner, N. J., L. C. Thompson, M. T. Thompson and A. Z. York. 1990. *Thompson Ethnobotany*. Memoir No. 3. Royal British Columbia Museum, Victoria.

Turner, Nancy J. 1995. *Food Plants of the Coastal First Peoples*. University of British Columbia Press, Vancouver.

———. 1997. *Food Plants of the Interior First Peoples*. Royal British Columbia Museum Handbook. University of British Columbia Press, Vancouver.

Vestal, P. A. [1952] 1973. *Ethnobotany of the Ramah Navaho*. Reports of the Ramah Project, Report No. 4. Reprinted in Kraus Reprint Company, Millwood, New York.

Westbrooks, Randy G., and James W. Preacher. 1986. *Poisonous Plants of Eastern North America*. University of South Carolina Press, Columbia.

Willard, T. 1992. *Edible and Medicinal Plants of the Rocky Mountains and Neighbouring Territories*. Wild Rose College of Natural Healing, Calgary.

PHOTO CREDITS

INDEX TO COMMON AND SCIENTIFIC NAMES

Page numbers in **boldface** type refer to the primary account headings.

INDEX

INDEX

INDEX

INDEX

INDEX

INDEX

INDEX

INDEX

ABOUT THE AUTHOR

*A*n avid naturalist since childhood, Linda finally focused on botany at the University of Waterloo, earning her master's degree in 1976. Following her education, she has worked as a consultant and researcher in northwestern Canada and as an editor/ author in Edmonton, while pursuing two favorite pastimes—photography and illustrating. Linda hopes that through her books people will glimpse some of the beauty and fascinating history of wild plants, and will recognize the intrinsic value of nature's rich mosaic.

More Great Books About the Outdoors!

Plants of the Rocky Mountains
By Linda Kershaw, Andy MacKinnon & Jim Pojar

Over 1200 plants from the Rocky Mountains, extending from Colorado, Wyoming, Montana and Idaho through the Canadian Rockies.

ISBN 1-55105-088-9 • 336 pages • over 800 color photographs, 1100 line drawings

Softcover $19.95 U.S. • $26.95 CDN

Mammals of the Rocky Mountains
By Chris Fisher, Don Pattie & Tamara Hartson

A colorful, illustrated field guide to 131 mammals common to the wide ranges of the Rocky Mountains.

ISBN 1-55105-211-3 • 296 pages • color photographs and illustrations

Softcover $18.95 U.S. • $26.95 CDN

Birds of the Rocky Mountains
By Chris Fisher

Over 320 common and interesting birds of the Rockies are brought to life by the colorful illustrations and detailed descriptive text.

ISBN 1-55105-091-9 • 336 pages • color illustrations

Softcover $19.95 U.S. • $24.95 CDN

Canadian Rockies Access Guide
By John Dodd & Gail Helgason

This essential guide for exploring the Rockies includes day hikes, backpacking, boating, camping, cycling, fishing and rainy-day activities.

ISBN 1-55105-176-1 • 400 pages • color photographs and illustrations

Softcover $16.95 U.S. • $19.95 CDN

Canadian Orders

1-800-661-9017 Phone

1-800-424-7173 Fax

LONE PINE

US Orders

1-800-518-3541 Phone

1-800-548-1169 Fax

E-mail: info@lonepinepublishing.com